Future Craftsmanship

Millennials & Gen Z in the Trades

by

Finn Westbrook

Future Craftsmanship

Millennials & Gen Z in the Trades

Contents

Future Craftsmanship

Millennials & Gen Z in the Trades

Contents

Chapter1: Crafting Tomorrow - The Role of Millennials & Gen Z

The world of craftsmanship is undergoing a transformative period, charged with the vibrant energy of Millennials and Gen Z. As we stride into the future, the fusion of tradition with technology unfolds a new chapter in the annals of making and crafting. This intersection is not just an evolution—it's a revolution, redefining the contours of creativity and innovation.

At the heart of this revolution are the young artisans and craftsmen of today, who wield their tools with an ethos that is markedly different from that of the generations before them. They are not just creators; they are innovators, educators, and visionaries, shaping a future where the value of handmade and bespoke does not dwindle but flourishes anew through the integration of technological advancements.

Millennials and Gen Z are not merely adopting the crafts of yesteryears as hobbies or professions. They are reimagining these crafts, embedding them with principles of sustainability, social responsibility, and digital efficiency. It's a testament to their vision that ancient techniques are now being melded with cutting-edge technologies, from 3D printing to virtual and augmented reality, creating not just products, but experiences.

The narrative that unfolds in the chapters of their craftsmanship is enriched by their unique values and visions for the future. They bring to the table a keenness for eco-friendly materials, a passion for learning through online platforms, and an entrepreneurial spirit that drives innovation in traditional trades. It's a blueprint for a future where craftsmanship not only thrives but leads the way in innovation and ethical practices.

This book aims to delve into the pivotal role that Millennials and Gen Z play in crafting tomorrow. By integrating technology and innovation into traditional trades, they offer a vision for a fulfilling and future-proof career in craftsmanship. It's a narrative that resonates with educators, industry professionals, and young individuals curious about the evolving landscape of craftsmanship.

Understanding the ethos of these generations is crucial. Defined by their relentless pursuit of purpose and authenticity, Millennials and Gen Z are rewriting the rules of the game. Their approach to craftsmanship is not just about the mastery of skills but about creating meaning and making a positive impact on the world. It's a holistic approach that encapsulates the spirit of modern craftsmanship.

The digital artisan emerging today is a figure who stands at the crossroads of technology and traditional craftsmanship. This new breed of craftsman leverages the power of the digital world to enhance the tactile beauty of the handmade. Through their lens, we can see the future of craftsmanship taking shape—a future where technology and handiwork coexist in harmony, each amplifying the beauty and potential of the other.

This journey into the future of craftsmanship is not devoid of challenges. Yet, it's precisely these obstacles that Millennials and Gen Z are adept at navigating. With an unquenchable thirst for knowledge and an innate ability to adapt, these generations are well-equipped to tackle the financial hurdles, the struggle for recognition, and the bureaucratic complexities that often accompany the life of a modern craftsman.

The ethos of sustainability rings loud and clear in the approaches taken by these young craftsmen. In a world increasingly aware of its environmental footprint, the shift towards eco-friendly materials and the renewal of upcycling practices are not just trends but necessities. Millennials and Gen Z stand at the forefront of this shift, championing sustainable practices in modern craftsmanship with zeal and innovation.

Education and skill development have also taken on new dimensions under the influence of these generations. The traditional apprenticeship model is being reimagined to fit into the digital age, with online learning platforms and virtual mentorship providing new pathways to mastery. It's a reflection of the changing landscape of education, where learning is lifelong, flexible, and accessible to all.

The digital age has also reshaped the economics of craftsmanship. With changed job markets and the rise of entrepreneurship, Millennials and Gen Z are finding new ways to monetize their crafts. They are not just makers; they are savvy businesspersons who understand the value of branding, social media, and community building in the success of their ventures.

The cultural impact of the DIY movement and the maker culture, significantly influenced by Generation Z, underscores a larger narrative of independence, creativity, and collaboration. Online communities and collaborative projects have become the new workshop, where ideas are shared, skills are honed, and innovations are born.

Moreover, the breaking of stereotypes, especially around gender roles in trades, marks another frontier where Millennials and Gen Z are leading change. Women in trades, equipped with skill, vision, and resilience, are not just participating but excelling and reshaping perceptions in what was traditionally a male-dominated sphere.

In "Crafting Tomorrow - The Role of Millennials & Gen Z", we stand at the threshold of this new era in craftsmanship. It's an era marked by the dynamism of youth, the promise of technology, and the enduring value of hands that create. As we venture into the following chapters, let us be inspired by the vision of these young craftsmen, for in their hands lies not just the craft of today, but the shape of tomorrow.

As we embark on this journey together, remember that the confluence of tradition and innovation in craftsmanship is not merely a trend; it is a testament to the human spirit's ceaseless quest for beauty, utility, and meaning. It is here, in the melding of the old with the new, that we find a hopeful vision for the future—a future crafted with care, precision, and boundless creativity by Millennials and Gen Z.

Chapter 2:
The Heritage of Craftsmanship: A Bridge to the Future

In an era where rapid technological advancements and digital transformations dominate our lives, the timeless value of craftsmanship emerges as an indispensable bridge to the future. This chapter delves into the heart of what crafts signify - not just as mere occupations, but as a profound expression of human creativity and skill passed down through generations. It's about understanding that the essence of craftsmanship lies in the meticulous attention to detail, the dedication to quality, and the passion infused in every piece created by hand. At its core, this conversation is an invitation to explore how we can preserve these rich traditions while embracing the innovations of today. We're standing at a critical juncture where the convergence of age-old techniques with revolutionary technologies presents an unprecedented opportunity for craftsmen around the globe. It's not just about keeping the tradition alive but redefining it, making it more relevant, accessible, and appealing to a generation that's on the brink of becoming the most influential force in the workforce. By weaving together the threads of tradition and innovation, we're not only paying homage to our heritage but also paving the way for a future where craftsmanship continues to thrive and adapt in a rapidly changing world. The journey ahead is filled with boundless possibilities, and it's up to us to ensure that the heritage of craftsmanship remains a vibrant and vital part of our collective future.

Understanding Traditions

In the tapestry of human development, traditions stand as the threads that connect generations, weaving the past into the present and future. Craftsmanship, with its deep roots in tradition, serves not only as a testament to human ingenuity and creativity but also as a bridge connecting us to our cultural heritage. This chapter delves into the essence of these traditions, highlighting why preserving them is pivotal for future generations.

Traditions in craftsmanship encapsulate more than just the methods and materials used; they embody the stories, values, and community spirit of their creators. As societies evolve, there's a growing concern that these ancient practices might fade into obscurity, overshadowed by the allure of modern technology and mass production. Yet, it's precisely within this juxtaposition that the true value of traditional craftsmanship begins to shine through.

Understanding traditions involves recognising their role in shaping identities. Across the globe, artisans carry forth the legacy of their ancestors, crafting items that are imbued with cultural significance. These items are not mere commodities but narrators of tales, each brushstroke, and chisel mark a testament to the artisan's dedication and the cultural heritage they represent.

The process of mastering traditional crafts is often a journey of personal and communal growth. Apprenticeships, where knowledge is passed down through generations, play a crucial role. This mentorship, rooted in patience and respect, transcends the simple transfer of skills; it fosters a deep connection between the master and apprentice, built on the foundational values of dedication, discipline, and respect for the material and the craft.

However, in an era dominated by rapid technological advancement and shifting economic landscapes, traditional crafts face significant challenges. The push for efficiency and scalability often sidelines these time-honoured practices. It's here that we must ask ourselves: what is lost when we eschew tradition for convenience?

The loss extends beyond the disappearance of unique crafts; it erodes the very fabric of our cultural identity and diversity. Each tradition that vanishes takes with it centuries of accumulated knowledge and a piece of our collective human story. Recognising this, there's a growing movement towards reviving and sustaining traditional crafts. This movement isn't driven by nostalgia but by an understanding of the intrinsic value these practices hold for educational, environmental, and socio-economic reasons.

Integrating traditional craftsmanship with modern innovation offers a pathway to preserving these practices. By adopting new technologies and materials, artisans can enhance the efficiency and appeal of their crafts without compromising on their core ethos. This fusion not only breathes new life into traditional crafts but also makes them relevant for contemporary society, providing a sustainable model for future generations to build upon.

Education plays a pivotal role in this preservation effort. By incorporating traditional crafts into the curriculum, we can ignite a passion for these practices in the hearts of young learners. This education isn't just about the techniques but also about understanding the cultural contexts and environmental considerations bound up in these practices.

Moreover, the global marketplace offers unprecedented opportunities for traditional craftspeople. Modern e-commerce platforms and social media can bring local traditions to a global audience, creating new economic pathways that were previously unimaginable. This visibility not only provides financial benefits but also fosters a greater appreciation of cultural diversity and interconnectivity.

But the journey doesn't stop at merely sustaining these traditions. The goal is to invigorate them, encouraging innovation within the frameworks of these time-honoured crafts. This innovation isn't about diluting the traditions but rather about ensuring their relevance and resonance with contemporary society.

Through the lens of craftsmanship, traditions remind us of our humanity — our ability to create, imagine, and connect across time and space. They teach us that progress doesn't have to come at the expense of our past. Instead, by understanding and embracing our traditions, we pave a path to a future where technology and tradition coexist, enriching each other.

This journey of understanding traditions is not just about preserving our past. It's about reimagining the future of craftsmanship. It's about offering a beacon of hope and inspiration for those who stand at the crossroads of tradition and innovation. As we move forward, let us carry the torch of our ancestors, illuminating the way for a new generation of craftsmen who are as dedicated to honouring the past as they are committed to building the future.

So, as we turn our gaze towards the horizon, let's not forget the foundations upon which we stand. The heritage of craftsmanship, with its rich tapestry of traditions, offers us a guiding light. It's up to us to ensure that this light burns brightly for generations to come, bridging the gap between the old and the new, the past and the future, tradition and innovation.

In conclusion, understanding traditions is not about dwelling in the past. It's about acknowledging the foundation they provide, celebrating their continued relevance, and fostering their evolution in the modern era. As we weave the future of craftsmanship, let's intertwine the threads of tradition with the strands of innovation, creating a fabric that is resilient, vibrant, and inclusive, ready to meet the challenges of the future head-on.

The Evolution of Trades

In the grand tapestry of human history, trades have always been at the heart of societal development and evolution. From the stone mason to the blacksmith, the skills and knowledge passed down through generations have been crucial to our progress. Yet, as we stand at the brink of what many are calling the fourth industrial revolution, marked by a fusion of technologies that is blurring the lines between the physical, digital, and biological spheres, the evolution of trades is accelerating at an unprecedented pace.

Historically, trades were defined by the mastery of manual skills and techniques developed over long years of apprenticeship and practice. These trades shaped the physical world around us, crafting everything from the homes we live in, to the clothes we wear, and the tools we use. The value of a craftsman was found in their hands-on expertise, precision, and the soul imbued into each of their creations.

However, the industrial revolution introduced machinery that changed the landscape of manufacturing, challenging the traditional role of craftsmen by automating processes that were once solely the domain of human hands. Yet, rather than diminishing the value of craftsmanship, this technological evolution sparked a transformative period of adaptation and growth in the trades.

Fast forward to today, we are witnessing another revolution. The rise of digital technology and automation tools is transforming the very nature of trades once again. This evolution is not just about the adoption of new tools; it's about how these tools are expanding the boundaries of what is possible within traditional trades, creating new forms of craftsmanship that integrate technology and innovation.

The modern tradesman is no longer confined to the workshop. They are digital artists, using software to design intricate pieces that can then be brought to life through 3D printing. They are innovators, applying technologies like augmented reality to envision projects in new and immersive ways. Even in fields as ancient as pottery or weaving, new materials and techniques are being introduced, marrying tradition with technology to create works of art that speak to both the past and the future.

This evolution is also reshaping the educational landscape of trades. Apprenticeships and trade schools are incorporating lessons on digital literacy, robotics, and sustainable practices to prepare the next generation of craftsmen for the challenges and opportunities of the 21st century. It's a bridge between the old and the new, ensuring that the heritage of craftsmanship is carried forward, but also adapted to meet the demands of a changing world.

Moreover, the evolution of trades is fostering a renewed sense of community and collaboration. Online platforms have opened up global networks where craftsmen from different disciplines and cultures can share ideas, techniques, and inspiration. This collective knowledge pool is driving innovation at an accelerated pace, making it an exciting time to be part of the world of trades.

One of the most profound changes in the evolution of trades is the blurring of lines between professions. Today's trades are not just about woodworking, metalworking, or textiles; they are interdisciplinary, combining skills and knowledge from various fields to innovate and create in ways that would have been unimaginable just a few decades ago.

But with all these changes, it's vital not to lose sight of the core values that have always underpinned craftsmanship: quality, durability, and aesthetics. The challenge for modern craftsmen is not just to embrace new technologies but to do so in a way that enhances, rather than detracts from, these fundamental principles.

This evolution also opens up new avenues for sustainability in trades. As environmental concerns become increasingly urgent, craftsmen are turning to eco-friendly materials and methods, driving a shift towards more sustainable practices within the industry. This alignment between innovation and sustainability underscores the potential for trades to contribute to solving some of the most pressing challenges of our times.

As we look to the future, the evolution of trades offers a vision of a world where tradition and innovation coexist harmoniously. It's a testament to the adaptability and resilience of craftsmen through the ages, continually finding new ways to express creativity and skill in the face of changing technologies and societal needs.

To educators, industry professionals, and young individuals looking to carve a path in the world of trades, this evolution presents an opportunity. It's a call to re-imagine what it means to be a craftsman in the 21st century, to explore the intersection of technology and tradition, and to forge a career that is not just about making a living but about continuing a legacy of innovation and excellence.

In embracing this evolution, we enable the heritage of craftsmanship to be not just a bridge to the past but a foundation for the future. It's about creating a world where the trades are not just preserved but are thriving, dynamic, and central to our progress as a society. This journey, with its challenges and opportunities, is not just about adapting to change but about leading it, shaping the future of trades with each creation that blends the best of tradition with the possibilities of tomorrow.

The journey of the evolution of trades is a mirror to our journey as a society – it tells a story of adaptation, resilience, and vision. It's a narrative that reaffirms the importance of craftsmanship not just as a means of livelihood but as a formative force in our cultural and technological evolution. It's a reminder that at the core of every innovation, every leap forward, there's a craftsman, bridging the past with the future, one creation at a time.

In conclusion, the evolution of trades is a vibrant, ongoing story of transformation. It's about honouring the roots of craftsmanship while reaching forward to new heights of creativity and innovation. For anyone with a passion for the craft, there has never been a more exciting time to be part of this story. Let's continue to build on this rich heritage, pushing the boundaries of what's possible, and together, create a future where the art of trade shines brighter than ever.

Chapter 3: Who are Millennials & Gen Z?

In the evolving narrative of craftsmanship, understanding the key players - Millennials and Gen Z - is essential. Born between the early 1980s and the mid-1990s, Millennials have witnessed the pre-digital era morph into a world dominated by technology. They've seen the advent of the internet, social media, and the smartphone revolution, placing them in a unique position where they value both traditional hands-on skills and the possibilities technology unveils. Gen Z, born from the mid-1990s to the early 2010s, has never known a world without the internet, making them even more digitally native. Yet, it's a mistake to view them only through their technological adeptness. Both generations are united by their search for meaning, sustainability, and authenticity in their work. They're not just looking for jobs; they're seeking careers that resonate with their values and offer a tangible impact on the world. This chapter delves into defining who Millennials and Gen Z are, outlining their core values and vision for the future. It's a closer look at these generations that are not just shaping the future of craftsmanship but are at the forefront of blending tradition with innovation to carve out fulfilling careers that honour the past while ushering in a new era of possibilities.

Defining the Generations

Before we delve deeper into the fusion of tradition and innovation within the realm of craftsmanship, it's essential to understand the protagonists of our narrative - the Millennials and Generation Z. These two cohorts stand on the cusp of a transformative era, bringing with them distinct perspectives, skills, and aspirations that are reshaping industries, including the traditional trades.

The term 'Millennial' commonly refers to individuals born between 1981 and 1996. This generation has witnessed the dawn of the digital age, experiencing first-hand the transition from analogue to digital. Their formative years were marked by significant technological advancements, which have undoubtedly influenced their approach to learning, work, and creativity. Millennials are often characterised by their adeptness at adapting to change, a trait that has enabled them to blend traditional craftsmanship with modern technology seamlessly.

On the other hand, Generation Z, those born from 1997 onwards, are true digital natives. From their earliest moments, this generation has been immersed in a world of high-speed internet, smartphones, and social media. This constant connection to a digital ecosystem has instilled in them an innate comfort with technology, along with a preference for visual and experiential forms of learning. Gen Z's insights into digital platforms are invaluable in redefining how artisanal trades are taught, practised, and marketed in the 21st century.

Both generations are entering or have entered the workforce with a fresh perspective on work-life balance, sustainability, and social responsibility. These values significantly influence their career choices, as many seek professions that not only provide personal fulfilment but also contribute positively to society. The traditional crafts, with their inherent emphasis on sustainability and personalised craftsmanship, are thus experiencing a resurgence of interest among these younger generations.

However, to attract and retain the interest of Millennials and Gen Z, craftsmanship must evolve. This evolution involves integrating digital technologies and sustainable practices into the core of trades.

Such a transformation not only respects the essence of traditional crafts but also makes them relevant and appealing to a generation that values innovation and environmental consciousness.

One of the most striking characteristics of Millennials and Gen Z is their entrepreneurial spirit. With access to a global market through the internet, many young craftsmen are exploring ways to turn their passion into viable businesses. This entrepreneurial inclination aligns perfectly with the nature of craftsmanship, which often thrives in a setting of individual creativity and small-scale production.

Moreover, the collective mindset of these generations, facilitated by digital platforms, fosters a culture of collaboration and sharing. This is evident in the rise of co-working spaces, online communities, and collaborative projects that bring together artisans from various disciplines. Such platforms not only serve as a crucible for innovation but also help in preserving and disseminating traditional techniques to a wider audience.

The emphasis on mental health and well-being is another area where the priorities of Millennials and Gen Z converge with the ethos of craftsmanship. The act of creating with one's hands is inherently therapeutic, offering a respite from the digital overload of modern life. As young individuals seek ways to counterbalance the stresses of their high-paced lives, the deliberate and contemplative nature of crafting offers a form of mindful escapism.

Education and skill development for these generations also require a departure from conventional methods. The availability of online resources and learning platforms has transformed the landscape of education, allowing for a more personalised and flexible approach to acquiring new skills. This digitalisation of education is particularly beneficial for the trades, where video tutorials and online workshops can provide accessible entry points for beginners.

The intersection of art and craftsmanship is another realm where Millennials and Gen Z are making significant contributions. Viewing craftsmanship not merely as a set of skills but as a form of artistic expression, they are challenging the boundaries between these domains. This perspective encourages a fusion of techniques and styles, promoting innovation while respecting traditional foundations.

In addressing climate change, both generations exhibit a strong inclination towards sustainable and eco-friendly practices. This sensitivity towards the environment is seamlessly woven into their approach to craftsmanship, from the choice of materials to the methods of production. The revival of upcycling and the exploration of sustainable materials are reflective of this ecological consciousness.

Despite their enthusiasm and potential to revitalise traditional trades, Millennials and Generation Z face significant challenges. These include financial hurdles, such as access to affordable apprenticeships and the high costs of setting up workshops. Moreover, societal perceptions that undervalue craftsmanship as a career path can also pose barriers to entry for young artisans.

In conclusion, Millennials and Generation Z are not just the torchbearers of traditional craftsmanship into the future; they are its potential transformers. Their unique blend of skills, values, and perspectives holds the promise of reintroducing craftsmanship to the modern world in a manner that is sustainable, innovative, and deeply rooted in humanistic values. As we look ahead, it's clear that the future of craftsmanship lies in embracing change while honouring tradition, a vision that these generations are uniquely equipped to realise.

For educators, industry professionals, and young individuals alike, understanding these generational distinctions and commonalities is crucial. It enables us to craft educational programs, collaborative platforms, and business models that not only cater to the needs and aspirations of these generations but also ensure the sustainability and relevance of craftsmanship in the years to come.

The journey of intertwining traditional trades with modern innovation is not without its challenges. However, by fostering an environment that encourages experimentation, supports learning, and celebrates creativity, we can inspire a new generation of craftsmen. These young artisans, equipped with the skills to navigate both the past and the future, are poised to redefine the landscape of craftsmanship, making it more vibrant, inclusive, and sustainable than ever before.

Values and Vision for the Future

At the heart of every craft lies a story of tradition, dedication, and innovation. As we delve into the aspirations and values of Millennials and Gen Z, it becomes clear that their vision for the future of craftsmanship is both a homage to the past and a blueprint for a sustainable, tech-integrated future. These generations are not just inheritors of traditional trades; they are pioneers at the forefront of a modern renaissance in craftsmanship.

The values that define Millennials and Generation Z are rooted in a blend of sustainability, creativity, and technology. They seek careers that not only provide financial stability but also align with their ethical beliefs and passions. The notion of crafting as a mere job is foreign to them; instead, they view it as a calling that is deeply intertwined with their identity and sense of purpose in the world.

This generation's vision is not limited by the confines of traditional workshops or the strictures of past methodologies. They are digital natives who see technology not as a threat to traditional craftsmanship but as a powerful ally. From 3D printing to virtual reality, they are embracing innovative tools to broaden the scope of what's possible in trades, turning them into art forms that engage with contemporary audiences.

Sustainability is another cornerstone of their value system. Millennials and Gen Z are acutely aware of the environmental impact of consumerism and production. Their approach to craftsmanship is guided by an ethos of eco-friendliness, favouring materials and processes that minimise waste and reduce carbon footprints. This commitment also drives the revival of upcycling practices, where the old is ingeniously transformed into the new, imbuing crafts with both history and innovation.

These younger generations also champion inclusivity and diversity in the world of crafts. They are breaking down barriers that once made certain trades seem exclusive to specific demographics. By fostering a culture of openness, they are welcoming voices that were previously underrepresented, thus enriching the craft community with a multitude of perspectives and styles.

The educational paths to becoming skilled in a craft have also evolved, thanks largely to the influence of these generations. Traditional apprenticeships are being reimagined to accommodate the realities of modern life, including online learning and hybrid models that offer flexibility without compromising on the depth of skill acquisition. This adaptation ensures that the transmission of knowledge remains uninterrupted, bridging generations.

Millennials and Gen Z are also redefining the notion of success in the realm of craftsmanship. For them, success is not only measured by financial gain but also by the impact of their work on the community and the environment. They are motivated by a desire to leave the world better than they found it, weaving social responsibility into the fabric of their creations.

Despite the challenges that come with pursuing a career in craftsmanship, including financial hurdles and the struggle for recognition, these young artisans remain undeterred. Their resilience is fueled by a belief in the value of their work and a vision for a future where craftsmanship is celebrated not just as a livelihood but as a vital expression of human creativity and sustainability.

The spirit of collaboration among Millennials and Gen Z is another defining feature of their approach to craftsmanship. They are leveraging online platforms to build communities where knowledge, skills, and resources are shared freely. This collective spirit is paving the way for interdisciplinary projects that merge various forms of creativity, offering new possibilities for innovation.

Social media has become a powerful tool for these generations, not just for marketing but for storytelling. They understand the power of narrative in connecting with their audience, using social platforms to share the journeys behind their crafts, thus fostering a deeper appreciation and demand for artisanal products.

Looking towards the future, Millennials and Gen Z are not content with merely adapting to the changing landscape of craftsmanship; they are determined to shape it. Their vision includes a world where traditional crafts are preserved while being dynamically integrated with new technologies and sustainable practices. They see the potential for craftsmanship to drive social change, promote mental well-being, and contribute to the economy in meaningful ways.

The challenge of balancing tradition and innovation is enthusiastically embraced by these generations. They recognise the value of historical techniques and are committed to preserving them, but not at the expense of progress. Instead, they are finding ways to honour the past while forging a future that reflects the evolving needs and values of society.

Ultimately, the vision of Millennials and Gen Z for the future of craftsmanship is one of hope, resilience, and transformation. They envisage a world where craftsmanship is not just a relic of the past but a living, breathing testament to human ingenuity. It's a future where crafts are not only valued for their aesthetic and utilitarian qualities but also for their contribution to a more sustainable, inclusive, and connected world.

As educators, industry professionals, and young individuals exploring the intersection of tradition and innovation, there's much to learn from the values and vision of Millennials and Gen Z. Their approach to craftsmanship offers a roadmap for a future where the craftsman's hand and the digital tool coexist in harmony, creating a legacy that generations to come will cherish and build upon.

Chapter 4:
The Digital Artisan: Technology Meets Handicraft

In the heart of the modern craft movement lies the concept of the digital artisan, a title that epitomises the fusion of cutting-edge technologies with traditional handcrafting methods. This chapter delves into the intriguing world where innovation enriches craftsmanship, rather than replaces it. Through the exploration of advancements such as 3D printing, virtual and augmented reality, and the Internet of Things (IoT), we'll uncover how these tools are not just reshaping the way artisans work but are also enhancing the creativity, efficiency, and reach of their crafts. The emergence of the digital artisan represents a pivotal shift in the landscape of craftsmanship, offering a beacon of possibility for those who yearn to merge their appreciation for the heritage of handiwork with the boundless potential of technology. This evolution paves the way for a future in which craftsmanship is not only preserved but also evolved, opening the doors to innovative creations that were once thought impossible. For educators, industry professionals, and the youth who stand at the brink of their career paths, the digital artisan serves as a compelling example of how one can embrace technological advancements while remaining deeply rooted in the time-honoured traditions of craftsmanship.

Innovations Reshaping Craftsmanship

In the heart of the modern renaissance of traditional crafts, a transformation is unfolding, driven by emerging technologies that are reshaping what it means to be an artisan in today's world. This fusion of technology and handicraft, not only preserves the ancient wisdom embedded in traditional practices but elevates it through precision and possibilities that were once deemed unfathomable. From laser cutting to 3D printing, these innovations are not just tools; they are collaborators that bring an artisan's vision to life with an accuracy and efficiency that complements human skill rather than replaces it. As we venture deeper into the age of the digital artisan, the boundless potential of integrating cutting-edge technologies with handmade crafts heralds a future where the tactile beauty of artisanal works and the infinite possibilities of digital innovation coalesce. This phenomenon is not merely a trend but a movement, affirming that in a world racing towards automation, the value of craftsmanship, imbued with the uniqueness of human touch and enriched by technology, stands more pronounced than ever.

3D Printing in Artisanship In the tapestry of modern craftsmanship, there's a thread that's increasingly becoming impossible to ignore: 3D printing. This technology, once the realm of engineers and big corporations, has started weaving its way into the artisan's toolkit, bringing with it a fusion of tradition and innovation that is reshaping the landscape of creation and design.

The introduction of 3D printing into artisanship represents a seismic shift not just in the way objects are made, but also in how creators approach the design process. For millennia, artisans have been limited by the constraints of their tools and materials. However, 3D printing technology offers a freedom that is unprecedented in the history of craftsmanship.

Imagine a sculptor, traditionally confined to the materials of clay, stone, or bronze, now able to experiment with durable, lightweight, and versatile plastics or resins. The possibilities for innovation in texture, form, and function are boundless, allowing artisans to push the boundaries of traditional sculpting.

Moreover, the integration of 3D printing in artisanship isn't just about creating new forms; it's about reviving old ones. There are artisans using 3D printing to restore and reimagine historical objects, breathing new life into artifacts that might otherwise crumble into obscurity. This act not only preserves our cultural heritage but also provides an invaluable learning resource for understanding ancient craftsmanship techniques.

However, the path to integrating 3D printing into traditional craftsmanship isn't without its challenges. There's a learning curve associated with mastering the technology—designers must become adept at using software to model their creations before they can be brought to life. Yet, this hurdle is also an opportunity for growth, prompting craftsmen to acquire new skills and adapt to the digital age.

One significant development spurred by 3D printing in artisanship is customisation. The technology allows for personalisation at a level that was previously difficult and costly to achieve. Artisans can now tailor their creations to the specific desires of their clientele, enhancing customer satisfaction and loyalty.

Environmental responsibility is another dimension of craftsmanship that is being transformed by 3D printing. With the ability to print on demand, waste is significantly reduced. Furthermore, the development of eco-friendly filaments means artisans can produce works that are not only beautiful but also sustainable.

The community aspect of craftsmanship is also evolving with 3D printing. Online platforms and forums have emerged as spaces where artisans can share their designs, techniques, and experiences. This digital collaboration expands the notion of the artisanal community beyond geographical limitations, fostering a global exchange of ideas.

Education in artisanship is leveraging 3D printing to provide more immersive and comprehensive training. Schools and workshops are beginning to incorporate this technology into their curricula, equipping the next generation of artisans with a skill set that marries the traditional with the futuristic.

In the realm of fashion and textiles, 3D printing is enabling designers to experiment with structures and materials that were previously impossible or impractical. This innovation has the potential to revolutionize the way we think about fabric, wearability, and function in fashion design.

The influence of 3D printing on artisanship extends to the culinary arts as well. Chefs and food artists are using the technology to create intricate designs and structures, pushing the boundaries of gastronomy and presentation. This fusion of taste and technology challenges our perceptions of what is possible in the culinary world.

Yet, embracing 3D printing in craftsmanship is not merely about adopting new technology; it's a mindset shift. It requires artisans to reconceptualise the act of creation, balancing their reverence for traditional methods with an openness to innovation.

As the fusion of 3D printing and artisanship continues to evolve, it promises not just a transformation of how things are made, but also a redefinition of what it means to be a craftsman in the modern world. It heralds a future where the only limit to creation is the artisan's imagination, where tradition and innovation walk hand in hand towards new horizons of design and beauty.

At its core, the integration of 3D printing into artisanship is a testimonial to the enduring human spirit of innovation. It's an invitation to all artisans, both seasoned and aspiring, to explore the boundless possibilities that this technology brings. Embracing 3D printing in craftsmanship isn't just about keeping pace with the times; it's about shaping the future with the tools of today.

In conclusion, the journey of 3D printing in artisanship is still unfolding, with each day offering new opportunities for growth, learning, and creation. For those willing to embrace these opportunities, the horizon is bright with potential. The fusion of technology and tradition is not just a trend; it's the blueprint for a future where craftsmanship continues to thrive, evolve, and enchant for generations to come.

Virtual Reality and Augmented Reality As we delve into the realms of virtual reality (VR) and augmented reality (AR), it's pivotal to understand how these technologies are revolutionising the landscape of craftsmanship and artisanship. These innovative tools aren't just reshaping how artisans work, but also how they conceptualise and visualise their creations, offering a bridge between traditional crafts and the digital future.

Virtual reality plunges the user into a completely digital environment, one that's fully immersive and interactive. For craftsmen, this means being able to simulate environments for testing designs, visualising complex structures in three dimensions before they've even laid hands on physical materials. It's a powerful tool that can enhance precision and creativity in equal measure.

Augmented reality, on the other hand, overlays digital information onto the real world, enhancing one's perception without creating a wholly separate environment. This infusion of digital and physical realms opens up new avenues for artisans. Imagine overlaying digital designs onto a piece of wood in real-time, aiding in intricate carvings that require a level of precision hard to achieve by eye alone.

The utility of AR in education cannot be overstated. For apprentices and students, it offers an interactive learning experience that transcends traditional classroom boundaries. Complex techniques can be demonstrated through augmented overlays, offering a hands-on learning experience without the material waste typically associated with trial and error.

One of the most enchanting aspects of VR and AR is their ability to preserve and archive traditional craftsmanship techniques. Through virtual environments, ancient crafts that are on the brink of extinction can be immortalised, allowing future generations to learn and experience these traditions in a deeply interactive way. This not only preserves cultural heritage but also inspires a new generation of artisans to explore old-world techniques through a modern lens.

Furthermore, VR and AR have the potential to democratise access to artisanal education. Geographic and socioeconomic barriers often limit access to traditional apprenticeships and craft education. Through virtual apprenticeships, aspiring craftsmen from any corner of the globe can learn from masters, without the need for physical presence. This opens up a world of opportunity for those who previously may have found the world of craftsmanship beyond their reach.

The integration of VR and AR in crafts also speaks to a broader narrative of sustainability. By enabling artisans to prototype designs virtually, there's a significant reduction in material waste. This aligns beautifully with the ethos of many modern craftsmen who champion sustainable and ethical practices within their work.

In marketing artisanal products, AR presents an innovative approach. Consumers can visualise products within their own space before making a purchase, reducing the uncertainty that sometimes accompanies buying bespoke items online. This not only enhances the customer experience but also bridges the gap between artisans and consumers, fostering a deeper connection to the craftsmanship behind the product.

Collaboration, an essential aspect of modern craftsmanship, is vastly enhanced by VR and AR. Artisans working in different locations can collaborate in a shared virtual space, discussing designs and making decisions in real-time as though they were in the same room. This collaborative potential can lead to innovative projects and products that might not have been possible otherwise.

However, the adoption of these technologies is not without challenges. There's a learning curve associated with mastering VR and AR tools, and the initial cost can be prohibitive for individual artisans or small workshops. Nonetheless, as these technologies become more mainstream and accessible, their transformative potential for the world of craftsmanship is undeniable.

The blend of VR and AR with traditional craftsmanship is not just about preserving the old with the new; it's about envisioning the future of trades. These digital tools empower artisans to push the boundaries of what's possible, encouraging innovation that respects tradition while boldly stepping into the future.

For educators in the field of craftsmanship, incorporating VR and AR into curriculums is an exciting prospect. It allows students to engage with materials and tools in a safe, controlled environment, fostering creativity and innovation from the outset. This approach also prepares the next generation of craftsmen for a future where digital and physical creations are intricately linked.

The potential of VR and AR to transform customer interactions is another area ripe with possibility. Beyond just visualising products, these technologies can offer immersive experiences that tell the story of the artisan's process, creating a deeper appreciation and understanding of the craftsmanship involved. This level of engagement can elevate artisan goods in the competitive global marketplace, highlighting the unique value they offer.

Lastly, it's vital for industry professionals and craftsmen to remain open to the possibilities that VR and AR present. Embracing these technologies can seem daunting, given the rapid pace of digital innovation. However, the benefits they offer in terms of education, collaboration, sustainability, and customer engagement make them crucial tools for the modern artisan. As we continue to explore the convergence of technology and traditional craftsmanship, VR and AR stand as beacons of innovation, illuminating the path forward for artisans around the world.

As we venture further into this digital renaissance of craftsmanship, the implications of VR and AR technologies are profound. They serve not just as tools for innovation, but as a means to connect more deeply with the heritage and future of craftsmanship. It's a journey of discovery, where ancient skills meet cutting-edge technology, forging a new era for artisans everywhere.

The Internet of Things (IoT) and Its Impact

In the realm of craftsmanship, where the tactile feel of material and the intimate knowledge of tools mark the passage of traditions, the introduction of the Internet of Things (IoT) stands as a beacon of change. This section explores the profound impact IoT technology has on the practices of modern artisans, transforming not only how objects are made but also how they're conceptualized and interacted with.

Iot, in essence, is about connectivity. It enables devices, tools, and materials to communicate with one another and with the internet in unprecedented ways. In the hands of a digital artisan, this connectivity can turn an ordinary object into a smart, interactive piece that enhances the user's experience while retaining the unique character of handmade craftsmanship.

For craftsmen and women, the IoT opens up a landscape rife with possibilities. Imagine a potter whose wheel can record the speed, pressure, and movements involved in creating a vase, then share that data with a community of learners. Or a carpenter whose handcrafted chairs can adjust their form based on the user's posture, learned over time through embedded sensors. This fusion of the digital and the tactile has the potential to elevate handicrafts into a realm where tradition meets innovation head-on.

The impact of IoT on traditional crafts does not stop at the creation process. It extends into the very essence of learning and passing on skills. Apprenticeship models, long the backbone of trade learning, can be revolutionised with IoT. Masters can monitor an apprentice's technique remotely, offering real-time feedback through connected devices. This not only bridges geographical gaps but also enhances the learning experience, blending the physical and digital worlds in skill acquisition.

Furthermore, IoT technologies help artisans in inventory management and supply chain logistics. By embedding sensors in materials, craftsmen can track usage, predict needs, and even automate the ordering process. This level of efficiency allows artisans to focus more on their craft and less on the mundanities of stock management.

The integration of IoT also opens up new avenues for artisans to engage with their audience. Consider a jeweller who creates pieces that change color or pattern based on the wearer's preferences, learned over time through connected apps. This personalisation creates a deeper bond between the maker and the user, transforming the artefact into an interactive experience.

However, integration with IoT is not without its challenges. The digital divide, cybersecurity, and the loss of privacy are valid concerns that artisans must navigate. While IoT offers immense potential, it requires a careful balancing act to ensure that the introduction of technology enhances rather than detracts from the essence of craftsmanship.

Moreover, the cost of implementing IoT technology can be prohibitive for small-scale artisans. Yet, as with all technological advances, costs are likely to decrease over time, making IoT more accessible to a broader range of craftsmen.

As traditional trades merge with digital technologies, environmental sustainability becomes increasingly crucial. IoT offers opportunities for craftsmen to monitor and reduce their environmental impact. Smart sensors can optimise energy use in workshops, and IoT data can help identify sustainable sources of materials.

The social impact of integrating IoT into craftsmanship cannot be overstated. It has the power to revitalise interest in traditional crafts, attracting a new generation drawn to the blending of tech and tactile. For young artisans, the allure of creating connected, interactive works can be the bridge that leads them to explore age-old practices.

Educationally, the introduction of IoT into craftsmanship creates a compelling case for STEAM (Science, Technology, Engineering, Arts, and Mathematics) learning. By illustrating the application of technology in arts and trades, educators can inspire students to pursue a broader range of interests and careers.

The global marketplace for handmade goods is also transformed by IoT. Artisans can reach a wider audience by offering customers the ability to customise products remotely, track the creation process, and even interact with the crafted objects in their homes. This global reach can significantly boost the viability and visibility of artisanal trades.

As digital artisans navigate this new landscape, the ethical considerations of replication and authenticity come to the fore. The uniqueness of a handcrafted piece lies in its imperfections and the story of its making. Though IoT can enhance the functionality and interaction of handmade items, artisans must tread carefully to retain the soul of their craft.

Ultimately, the integration of the Internet of Things into handicrafts presents a fascinating paradox. It challenges craftsmen to re-evaluate the boundaries of their art, urging them to embrace new technologies while holding steadfast to the principles of craftsmanship. This dynamic tension between the past and the present, the digital and the physical, is where the future of craftsmanship is being forged.

In conclusion, the intersection of IoT and traditional craftsmanship heralds an era of unprecedented innovation. As artisans blend these worlds, they not only preserve the heritage of their trades but also chart a course for a future where craftsmanship continues to be a vital, relevant, and evolving form of expression. The digital artisan, thus, stands at the threshold of a new dawn, where the melding of technology and tradition crafts a legacy for the generations to come.

Chapter 5:
Sustainable Practices in Modern Craftsmanship

In the journey through modern craftsmanship, we find ourselves at a crucial crossroads, where the path laid by generations before us intersects with innovative approaches geared towards sustainability. Chapter 4 delves into 'Sustainable Practices in Modern Craftsmanship', highlighting how today's artisans are gracefully weaving eco-friendliness into the fabric of their endeavours. Anchored in a deep respect for nature and an understanding of our environmental responsibilities, these craftspeople are not merely creating; they're stewarding the planet for future generations. By sourcing materials thoughtfully, rediscovering the magic of upcycling, and embracing techniques that leave a lighter footprint on the earth, they're moulding a world where craftsmanship doesn't just tell a story of culture and skill but also one of conscious conservation. This chapter lays bare the truth that sustainability in craftsmanship isn't just about the end product, but about the entire process - from the drawing board to the hands of the consumer. As we move forward, it's clear that the fusion of innovation and tradition in crafting sustainable solutions not only enriches our present but secures a legacy of care, excellence, and environmental stewardship for the artisans of tomorrow.

Eco-Friendly Materials and Methods

As we delve into the realm of sustainable practices in modern craftsmanship, it's imperative to understand the significance of eco-friendly materials and methods. The surge in environmental awareness has catalysed a seismic shift towards sustainability, echoing the ethos of newer generations who prioritise the health of our planet. Integrating eco-friendly materials and methods isn't just a trend; it's a testament to the collective responsibility we share towards fostering a sustainable future.

One of the cornerstone materials in sustainable craftsmanship is reclaimed wood. The allure of repurposed timber lies not only in its environmental benefits but also in its aesthetic and historical significance. Each piece tells a unique story, carrying the essence of its past life into its new purpose. Craftsmen who embrace reclaimed wood contribute to reducing deforestation and landfill waste, embodying the principles of responsibility and conservation.

Bamboo, hailed as a miracle of nature, stands out for its rapid growth and versatility. This swiftly renewable resource has found its place in the arsenal of eco-conscious craftsmen worldwide. From furniture making to architectural applications, bamboo's strength and lightweight characteristics make it a preferred choice for sustainable projects. Its cultivation requires no fertilisers or pesticides, making it a beacon of ecological harmony.

In the textile industry, the shift towards organic cotton and hemp reflects a growing awareness of the environmental impact of conventional cotton farming. Organic cotton thrives without harmful chemicals, safeguarding both the land and the farmers. Hemp, with its low water requirement and absence of pesticide dependency, represents a sustainable alternative with a minimal ecological footprint. These materials symbolise a commitment to cleanliness, quality, and sustainability in craftsmanship.

The use of natural dyes in crafting paints a vibrant picture of sustainability. Synthetic dyes, with their hazardous chemicals, pose significant environmental threats. In contrast, natural dyes offer a palette of rich colours derived from plants, minerals, and even insects, reducing pollution and fostering a healthier ecosystem. Craftsmen exploring natural dyes embark on a journey of discovery, reconnecting with ancient methods that celebrate the earth's bounty.

The methodology behind sustainable craftsmanship extends beyond material selection. Traditional techniques, honed over centuries, embody the essence of environmental stewardship. Hand tools, for instance, require no electricity, minimising carbon footprint. They represent a harmonious balance between human skill and ecological mindfulness, reminding us of the value inherent in taking time to produce less but of higher quality.

Upcycling, a concept that has gained considerable traction, showcases the transformative power of creativity in extending the life cycle of materials. By reimagining the potential of discarded items, craftsmen can produce works of art and functionality that challenge the throwaway culture. This not only reduces waste but also highlights the potential for innovation within the constraints of sustainability.

Water usage and waste management constitute critical considerations in crafting practices. Techniques that minimise water use and recycle waste products stand at the forefront of sustainable craftsmanship. By adopting such methods, craftsmen advocate for a holistic approach to sustainability, recognising the interconnectedness of resources and the imperative to preserve them.

The role of digital tools in sustainable craftsmanship cannot be overlooked. From 3D printing using recycled plastics to computer-aided design (CAD) minimizing material waste, technology offers innovative solutions to age-old challenges. These digital methods furnish craftsmen with the tools to design and produce with precision, efficiency, and a reduced environmental impact.

Energy efficiency in crafting processes also holds immense potential for sustainability. Renewable energy sources, such as solar and wind power, are making inroads into workshops and studios around the world. By harnessing these clean energy sources, craftsmen can mitigate their carbon footprint, fostering an environment of true sustainability.

The adoption of eco-friendly packaging and shipping methods further exemplifies the comprehensive approach required in modern craftsmanship. Biodegradable and recycled materials for packaging, coupled with carbon-neutral shipping options, complete the cycle of sustainability, ensuring that every step in the process aligns with ecological principles.

Educating clients and consumers plays a pivotal role in advancing the agenda of sustainable craftsmanship. Awareness campaigns and transparent communication about the environmental benefits of eco-friendly products can shift consumer preferences towards sustainability. By making informed choices, consumers become active participants in the eco-friendly movement, driving demand for sustainable crafts.

Partnerships between craftsmen and environmental organizations offer another avenue for promoting sustainability. These collaborations can lead to innovative projects and initiatives that highlight the importance of eco-friendly materials and methods, inspiring communities to embrace sustainable practices.

Acknowledging the challenges in adopting sustainable practices is crucial. The initial costs and time involved in sourcing eco-friendly materials or mastering traditional techniques can be deterrents. However, the long-term benefits — environmental preservation, reduction of waste, and the creation of healthier living spaces — make these endeavours worthwhile.

As we forge ahead, the integration of eco-friendly materials and methods in modern craftsmanship embodies a vision of hope and resilience. It underscores a commitment to not only preserving our planet but also enriching our lives through sustainable practices. Craftsmen stand at the vanguard of this movement, wielding the tools of innovation and tradition to build a more sustainable, aesthetically enriching world.

The Renewal of Upcycling

Sustainable practices are becoming increasingly pivotal in the world of modern craftsmanship, providing a roadmap for environmentally conscious creation. In the essence of sustainability comes the practice of upcycling, an age-old concept that's been imbued with new life and relevance in our rapidly changing world. Upcycling, unlike recycling, doesn't just aim to reuse materials. It elevates the value of what would otherwise be considered waste, transforming it into items of even greater worth and utility. This practice not only reduces the strain on our planet's resources but also challenges craftsmen to innovate and reinvent.

The modern artisan's workbench has evolved, with upcycling taking centre stage. It's not merely about breathing new life into discarded materials; it's a statement of intent, a demonstration of respect for the environment, and a tangible step towards reducing the craftsmanship sector's carbon footprint. By reusing materials, craftsmen cut down on the energy-intensive processes of production, such as extraction and manufacturing, directly contributing to the conservation of the environment.

Upcycling possesses a dual allure. For one, it encapsulates the very essence of creativity, compelling craftsmen to reimagine the utility and aesthetics of materials. It requires a vision that sees beyond the current form and perceives the potential locked within. This level of creativity can catalyse a new wave of design thinking, spurring innovation that, while grounded in tradition, is not confined by the limitations of the past.

The other facet of upcycling's appeal is its economic incentive. In a market where consumers are increasingly seeking unique, eco-friendly products, upcycled goods stand out. They offer individuality and story, traits that are becoming more valuable than ever in our mass-produced world. These items carry narratives of transformation and redemption, which resonate with a consumer base that is becoming more conscientious about the origin and life-cycle of the products they purchase.

An emblematic shift can be seen in the furniture industry, where upcycling has ignited a resurgence of interest in vintage pieces. These items are not simply restored, but reimagined, often incorporating modern design elements or functionality. This fusion of old and new not only satisfies the contemporary eye but also pays homage to the workmanship of the past, ensuring that these skills and designs are not forgotten but instead, celebrated and repurposed.

Fashion, too, has taken note of the upcycling revolution. Old garments, textiles, and even plastic materials have found their way onto runways as part of cutting-edge collections. Designers are deconstructing and reconstructing materials, combining them with innovative techniques to create fashion that's at once both avant-garde and deeply responsible.

The upcycling movement has also been a boon for community engagement and education. Workshops and classes designed to teach the art of upcycling have sprung up globally, serving as incubators for new ideas while fostering a sense of communal responsibility for the environment. These spaces become local hubs where skills are shared, and sustainability is practised in a tangible, hands-on manner.

Within the ethos of upcycling, there exists a narrative that speaks to the core of human resourcefulness. From turning pallets into elegant furniture to crafting jewelry from electronic waste, these are stories of innovation that do not merely seek to solve a problem, but to spin the problem itself into something magnificent.

Traditional trades such as metalworking and woodworking are discovering new dimensions within upcycling. They are not simply repeating history but are redefining the future of their crafts. Metal artisans, for instance, are transforming scrap into striking sculptures, and woodworkers are using reclaimed wood to create pieces that are as poignant as they are elegant.

Upcycling also represents a responsible approach to globalisation. As craftsmen from different parts of the world exchange ideas, they discover universal methods for reusing materials that were once seen as mere waste. This global conversation not only spreads the movement but also leads to innovation driven by diverse cultural approaches to materials and design.

This practice challenges the craftsman's dedication to their trade, as it demands constant learning and adaptation to work with materials that were never intended for a second life. It teaches patience, innovation, and a deep understanding of material properties. In many ways, upcycling brings craftsmen back to the core of their craft: the ability to create something extraordinary from the ordinary.

As a form of art, upcycling has found its way into galleries and exhibitions, challenging the public's perception of what constitutes art and what defines waste. These displays are not merely pieces to be admired but are conversations starters about consumption, waste, and our responsibility toward future generations.

In conclusion, upcycling is more than just a trend; it's a renewal of the ethos of craftsmanship. It's a way for modern craftsmen to honour the materials, tools, and techniques of their craft, while simultaneously addressing the pressing issue of environmental sustainability. Through invention and reinvention, upcycling shows us that by looking backward with respect and forward with optimism, we can forge a future that is as sustainable as it is inspired.

The conversation surrounding upcycling in the context of modern craftsmanship is a profound one, acknowledging the responsibility of creators to not only generate beauty and functionality but also to act as stewards of the environment. The renewal of upcycling is a testament to the resilience and adaptability of craftsmen, proving that even in a technologically advanced age, there is infinite value in the reimagining of the materials that the Earth provides us.

Chapter 6: Education and Skill Development

As we delve into the heart of education and skill development, it's clear that for craftsmanship to thrive in the modern age, a reinvention of learning methods is essential. Apprenticeships and trade schools have long been the backbone of passing down skills from one generation to the next. However, as we stand at the junction of tradition and innovation, it's evident that these institutions must evolve. The adaptation of online learning platforms within trade schools is not just a nod towards modernity but a necessity to keep pace with the changing landscape of the industry. Embracing digital tools and technologies within education doesn't dilute the essence of craftsmanship; rather, it enriches it, allowing for a blend of precision, creativity, and efficiency previously unimaginable. This chapter doesn't just argue for a mere integration of technology into educational models but advocates for a holistic reimagining of how we view, value, and undertake skill development in the 21st century. It aims to serve as a blueprint for educators, industry professionals, and young individuals, illustrating that the path to mastering a trade isn't confined to the walls of traditional learning but is expansive and brimming with potential if we dare to innovate how we learn as much as what we learn.

Reimagining Apprenticeships

In a world where traditional pathways are being constantly reevaluated in the light of modern demands and technologies, the concept of apprenticeships—a cornerstone in the realm of craftsmanship and skill development—warrants a fresh perspective. With the evolving landscape, there's an exciting opportunity to blend age-old wisdom with the possibilities offered by the digital era.

Historically, apprenticeships have been the backbone of learning trades, offering hands-on experience and mentorship. Yet, as we plunge deeper into the 21st century, it's clear that the traditional model needs a revision to stay relevant. The integration of technologies such as 3D printing, virtual reality, and augmented reality into the apprenticeship curriculum can transform the learning experience, making it more immersive, efficient, and appealing to a generation that's grown up in a digital world.

Moreover, there's a growing recognition of the diverse learning styles and needs of individuals. Personalised learning paths enabled by technology can provide apprentices with a more tailored experience, ensuring that they acquire the skills in a manner that best suits their learning preferences. This personalised approach not only enhances skill acquisition but also increases engagement and motivation among learners.

Collaboration and knowledge sharing have been amplified by the internet, breaking down geographical barriers and creating global communities of craftsmen and apprentices. Imagine leveraging this connectivity to facilitate virtual apprenticeships where learners have the opportunity to work with mentors from around the globe. This not only enriches the learning experience but also fosters a deeper understanding and appreciation of global craftsmanship traditions and innovations.

Environmental sustainability has become a critical concern in all sectors, including craftsmanship. Reimagined apprenticeships can play a pivotal role in promoting eco-friendly practices. By integrating sustainable materials and methods into the curriculum, apprentices can be equipped with the knowledge and skills needed to contribute to a more sustainable future in their respective trades.

Entrepreneurial skills are increasingly important in today's economy. Many craftsmen dream of starting their own ventures. Incorporating business and entrepreneurial education into apprenticeship programmes can empower the next generation of craftsmen with the skills needed to navigate the market, brand their craft, and build successful businesses.

The fusion of craftsmanship and digital technology opens up new avenues for innovation. Apprentices should be encouraged to experiment with combining traditional techniques with modern technologies, fostering a culture of innovation and creativity. This approach not only leads to the development of unique products and solutions but also ensures that traditional crafts evolve and remain relevant in the modern world.

Mental health and wellbeing have become forefront concerns, particularly amongst the younger generations. An apprenticeship model that emphasises a supportive community, encourages mindfulness, and integrates practices that enhance mental well-being can make the learning experience more holistic and fulfilling.

Accessibility and inclusivity are paramount to ensure that apprenticeships are available to everyone, regardless of their background, abilities, or circumstances. This means not only breaking down physical barriers but also challenging stereotypes and biases that might deter individuals from pursuing certain trades. An inclusive approach to apprenticeship can help to diversify the trades, bringing in fresh perspectives and talents.

Assessment methods within apprenticeships also warrant rethinking. Moving away from purely outcome-based assessments to a more holistic approach, which takes into consideration the apprentice's growth, creativity, problem-solving skills, and ability to innovate, can provide a more accurate measure of their capabilities and readiness to embark on their professional journey.

The role of apprenticeships in career progression should also be highlighted and expanded. They should not be seen merely as a stepping stone but as a continuous learning pathway that offers opportunities for advancement, specialisation, and mastery within a craft. This perspective encourages lifelong learning and fosters a deeper commitment to one's trade.

Collaborative projects and interdisciplinary learning should be incorporated into apprenticeship programmes. By working on projects that require a combination of skills and collaboration with individuals from different trades or disciplines, apprentices can learn the value of teamwork, diversify their skills, and adopt a more innovative approach to problem-solving.

Finally, the success of reimagined apprenticeships relies on the commitment of educators, industry professionals, and policymakers to support and invest in these programmes. It calls for a collective effort to ensure that apprenticeships remain a viable and attractive pathway for skill development and that they adequately prepare individuals for the challenges and opportunities of the future.

As we stand at the crossroads of tradition and innovation, reimagining apprenticeships presents a unique opportunity to revitalise craftsmanship for the 21st century. It's about bridging the past with the future, honouring the legacy of the trades while making them relevant and enticing for the new generation. This is not just a transformation of the apprenticeship model but a renaissance of craftsmanship itself, promising a future where skill, creativity, and technology converge to create the extraordinary.

In an ever-changing world, the reimagining of apprenticeships is not merely an option but a necessity. It's a call to action for all stakeholders involved in craftsmanship and education to come together and shape a future that values skills, nurtures talent, and celebrates innovation. By embracing change and redefining apprenticeships, we can ensure that the crafts that have enriched our past will continue to flourish and inspire future generations.

Trade Schools in the 21st Century

The landscape of education and skill development is undergoing a remarkable transformation, especially within the realm of trade schools, now that we've stepped into the 21st century. As we navigate through this era, the integration of cutting-edge technology and innovative methodologies in trade learning environments is becoming increasingly prominent. These institutions are no longer strictly places where manual skills are honed; instead, they've evolved into vibrant hubs where traditional trades and modern technological advancements converge. This evolution is pivotal in preparing a new generation of craftsmen, who are not just skilled with their hands but are also adept at leveraging digital tools to enhance their craft. The goal is to foster a workforce that's versatile, adaptive, and future-proof, capable of meeting the changing demands of the global economy. Trade schools are thus redefining themselves as essential pillars of education and skill development, seamlessly blending the rich heritage of craftsmanship with the endless possibilities presented by the digital age. By doing so, they're laying a solid foundation for a future where tradition and innovation coexist, offering young individuals and educators alike a new vision for what a fulfilling career in the trades can look like.

Online Learning Platforms As we progress through this journey of integrating traditional craftsmanship with the innovations of the digital age, it's essential to highlight the role of online learning platforms. These platforms aren't just supplementary tools; they are revolutionizing how we approach education in the trades. They democratize access to knowledge, break down geographical barriers, and offer a plethora of resources tailored to diverse learning styles.

The essence of modern craftsmanship lies in its ability to adapt and evolve. Online learning platforms serve as a bridge between generations, offering a space where both seasoned artisans and beginners can learn from each other. These platforms host a vast array of courses, covering everything from basic woodworking techniques to the complexities of metal forging, all at the click of a mouse.

One of the key advantages of these platforms is their flexibility. Learners can set their own pace, choosing when and where they study. This is particularly beneficial for millennials and Gen Z, who often juggle multiple responsibilities and prefer learning environments that can adapt to their hectic schedules.

Interactive formats, such as video tutorials, webinars, and forums, enrich the learning experience on these platforms. They allow learners to see techniques in action, ask questions in real-time, and engage in discussions with a global community of like-minded individuals. This interactive approach not only enhances skill acquisition but also fosters a sense of belonging and community among learners.

Moreover, online learning platforms often integrate elements of gamification, which can significantly increase motivation and engagement. By earning badges, certificates, or completing milestones, learners can track their progress and stay motivated through tangible achievements.

The affordability of these platforms is another factor not to be overlooked. They often offer free courses or charge a fraction of the cost of traditional education methods. This opens up possibilities for many who might not have the financial means to attend a trade school or undertake a formal apprenticeship.

It's also worth noting the role of these platforms in perpetuating the heritage of craftsmanship. Through digital archives, interviews with master craftsmen, and historical documents, learners can access a wealth of knowledge that was previously confined to libraries or specific locations.

However, the benefits of online learning platforms go beyond just skill acquisition. They encourage a culture of continuous learning and curiosity. In a world where technology and trends rapidly evolve, the ability to adapt and learn new skills is invaluable. These platforms cultivate a mindset that is not only beneficial for individual craftsmen but essential for the continuation and evolution of the trades themselves.

Yet, despite these advantages, the effectiveness of online learning varies among individuals. It requires discipline, self-motivation, and a proactive approach to learning. Therefore, these platforms often incorporate community support systems, mentorship programmes, and personalised learning paths to help learners navigate these challenges.

The integration of virtual and augmented reality technologies is also enhancing the learning experience on these platforms. These technologies offer immersive learning experiences that can simulate real-life scenarios, providing learners with a deeper understanding of the materials and techniques discussed. This is particularly beneficial for complex or hazardous trades that require a high level of precision and safety.

Furthermore, online platforms provide a gateway to entrepreneurship and innovation within the trades. They offer courses on business skills, marketing, and branding, enabling craftsmen to turn their passions into viable career paths. This holistic approach to education is shaping a new generation of digital artisans who are not only skilled in their craft but also adept at navigating the challenges of the modern business world.

In conclusion, the emergence of online learning platforms is a testament to the resilience and adaptability of craftsmanship in the digital age. These platforms are not just changing how knowledge is shared; they are changing the very fabric of education in the trades. They offer a promising avenue for preserving traditional skills, fostering innovation, and inspiring a new generation of craftsmen.

As we move forward, it's imperative that we continue to support and develop these platforms. By doing so, we ensure that the rich heritage of craftsmanship is not only preserved but also enriched with the possibilities of the digital age. Let us embrace these tools with enthusiasm and optimism, for they hold the key to a future where tradition and innovation coexist in harmony.

In essence, online learning platforms are more than just educational tools; they are catalysts for change. They represent a shift in how we perceive education and craftsmanship, shining a light on the vast potential that lies in the intersection of technology and tradition. As we journey into this promising future, let us not forget the core principles of craftsmanship: quality, ingenuity, and a relentless pursuit of excellence. Guided by these values and empowered by technology, there is no limit to what we can achieve.

Chapter 7: The Economics of Craftsmanship in the Digital Age

In an era where the digital and physical realms increasingly intertwine, the economic landscape of craftsmanship is undergoing profound transformations. As we delve deeper into this pivotal chapter, it becomes evident that traditional vocations, once perceived as endangered, are finding new vitality through digital augmentation. This resurgence is not merely a tale of survival but of thriving innovation and entrepreneurship that challenge outdated notions of what it means to be a craftsman in the 21st century. The modern artisan now navigates a dynamic job market where skills in digital tools and social media marketing are as crucial as the dexterity of their hands. These changes are paving the way for a future where craftsmanship and technology coexist in a symbiotic relationship, empowering artisans to reach wider markets and create with greater efficiency and impact.

The emergence of platforms for e-commerce, social media, and crowdfunding has revolutionized how craftsmen connect with their audience, turning local trades into global enterprises. This digital renaissance, while introducing new challenges, unlocks unprecedented opportunities for innovation, collaboration, and entrepreneurship. Artisans are no longer confined by the walls of their workshops; they are now part of a vibrant, interconnected global economy that values uniqueness and sustainability. This chapter explores how embracing modern technologies and business models can not only generate economic vitality but also secure craftsmanship's place in the future economy—as a thriving bastion of culture, innovation, and resilience. It's a call to action for the next generation of craftsmen to harness the potential of the digital age, ensuring their crafts not only endure but flourish in the ever-evolving economic landscape.

Changing Job Markets

The job market, as we've known it, is undergoing profound transformations, moulding under the pressures of technological advancements and shifting societal values. This change, while daunting for some, presents an unmatched opportunity for those in the realm of craftsmanship. Embracing technology and innovation within traditional trades is not just an option; it's becoming a necessity to thrive in the contemporary job landscape.

Today, the demand for personalised, high-quality products and experiences is surging. Consumers are increasingly valuing the story behind what they purchase, favouring items that carry a unique touch, a human element that mass production can't replicate. This shift has rekindled interest in craftsmanship, propelling it from the peripheries of the economy to the centre stage of the digital age.

However, this renewed interest comes with its set of expectations. The modern consumer isn't just looking for the traditional; they crave the innovative fusion of age-old skills with modern technology. Digital tools and platforms have become the new workbenches and marketplaces for today's craftsmen. From online sales platforms to social media marketing, digital fluency is becoming as essential as the mastery of traditional techniques.

Moreover, the proliferation of digital fabrication technologies, such as 3D printing and CNC machining, has broadened the horizon for what can be crafted. These technologies enable artisans to create with precision and efficiency, previously unimaginable, opening up new avenues for innovation within trades. Yet, the adoption of these tools doesn't undermine the value of handmade; rather, it enhances the artisan's capacity to customise and innovate, retaining the soulfulness of their work.

Simultaneously, sustainability has surged to the forefront of consumer consciousness, prompting a reevaluation of materials and methods in crafting. Today's craftsmen are expected to not only be adept at their trade but also be stewards of environmentally friendly practices. This emphasis on sustainability is redefining vocational success, aligning it with the broader goal of societal wellbeing.

Education and skill development in this evolving landscape are witnessing a transformation as well. Traditional apprenticeship models are being supplemented with online platforms and virtual reality tools, offering new ways to learn and master a craft. This accessibility to knowledge and training is democratising craftsmanship, enabling a wider range of individuals to partake in and contribute to the trade.

These changes in the job market are indicative of a broader societal shift towards valuing creativity, sustainability, and personalisation. This shift is not just expanding the scope of opportunities for craftsmen but is also elevating the societal perception of trades, recognising them as essential contributors to the economy and culture.

However, navigating this changing landscape requires more than just adaptability. It demands a reimagining of the role of craftsmen in society. They are no longer just the creators of tangible artifacts; they are innovators, educators, and custodians of culture. Understanding the economics of craftsmanship in the digital age means recognising the multifaceted value that craftsmen bring to the table - economic, cultural, and environmental.

The challenge for today's and tomorrow's craftsmen is to maintain the delicate balance between tradition and innovation. While technology offers tools to expand and enhance their craft, the essence of craftsmanship - the human touch, the story, and the connection - remains central. It's this essence that endears craftsmanship to people's hearts and lives, ensuring its relevance and resilience in the digital age.

Furthermore, the global nature of the digital economy offers craftsmen an unprecedented opportunity to reach international markets. The digital age has dismantled geographic barriers, allowing local artisans to share their work with a global audience. This international exposure not only opens economic opportunities but also fosters cultural exchange, enriching the global craft community.

In conclusion, the changing job markets present a tapestry of challenges and opportunities for craftsmanship in the digital age. Embracing technology, innovating within traditions, prioritising sustainability, and leveraging the global digital marketplace are not just pathways to economic success but are also avenues for craftsmen to contribute meaningfully to society's evolving narrative. As the job market continues to transform, the resilience and relevance of craftsmanship will hinge on its ability to adapt, innovate, and inspire.

The integration of craftsmanship with digital technology isn't a departure from tradition; it's an evolution of the trade, ensuring its survival and prosperity in the modern era. As we navigate these changing waters, it's crucial to foster a community that supports lifelong learning, creativity, and collaboration. Through this community, craftsmen can continue to build upon the rich heritage of their trades, while shaping them to meet the needs and desires of the future.

Ultimately, the future of craftsmanship in the digital age rests on the shoulders of those willing to navigate the changing tides. With the right blend of tradition, innovation, and sustainability, the craftsmanship sector can not only survive but thrive, offering meaningful, fulfilling careers to future generations. The changing job markets, therefore, are not a threat but a beacon, guiding the way towards a vibrant and sustainable future for craftsmanship.

Entrepreneurship and Innovation

In the tapestry of modern craftsmanship, the threads of entrepreneurship and innovation are interwoven with a finesse that marks a new era. This merging doesn't just highlight the evolution of craft; it underscores a radical rethinking of what it means to be a craftsman in the digital age. Gone are the days when artisans were confined to workshops, their talents tucked away in the folds of obscurity. Today's craftsmen are entrepreneurs, innovators, and visionaries, transforming their passions into thriving businesses that resonate with the ethos of the 21st century.

The digital age has democratised access to markets in a way previously unimaginable. E-commerce platforms and social media have become the modern craftsman's marketplace, enabling them to reach a global audience from their local studio. However, this broadened access comes with its own challenges. The market is more crowded than ever, and standing out requires not just skill in one's craft, but a savvy understanding of branding, marketing, and the digital landscape.

Entrepreneurship in craftsmanship today is not simply about making and selling. It's about telling a story, creating an experience, and building a community. Consumers are looking for authenticity and a connection to the makers of the items they purchase. They want to know the stories behind the products, the names and faces of the artisans. This shift towards storytelling and authenticity opens up new avenues for craftsmen to innovate and differentiate their brand.

Moreover, innovation is not merely technological; it's conceptual. The burgeoning field of sustainable and eco-friendly practices is an area ripe for innovation. Artisans are experimenting with recycled materials, developing new methods to reduce waste and decrease their environmental impact. By integrating these practices into their businesses, craftsmen are not only responding to consumer demand for sustainability but are also setting new standards within their industries.

Of course, the journey of marrying craftsmanship with entrepreneurship and innovation is strewn with challenges. Many artisans struggle with the transition from maker to entrepreneur, finding it difficult to navigate the realms of business operations, funding, and digital marketing. This struggle underscores the importance of education and skills development in entrepreneurship for craftsmen, framing it not just as a necessity but as a cornerstone of modern craftsmanship.

One emerging trend addressing these challenges is the rise of collaborative spaces and maker communities. These hubs not only provide shared resources and equipment, reducing the entry barriers for emerging craftsmen but also foster a culture of collaboration. Within these communities, knowledge sharing and mentorship are invaluable, allowing seasoned entrepreneurs to guide novices, bridging gaps in business acumen with craftsmanship expertise.

The role of technology in this paradigm cannot be overstated. From the advent of 3D printing to the use of augmented reality for design and presentation, technology offers craftsmen unprecedented tools to innovate and create. These tools are not replacing traditional techniques but rather augmenting them, allowing artisans to push the boundaries of their craft.

Yet, embracing technology and entrepreneurship requires a mindset shift. Craftsmen must see themselves as innovators, willing to experiment and take risks. This mindset shift is crucial not only for individual success but for the survival and evolution of craftsmanship itself. As industries evolve and new technologies emerge, the crafts that thrive will be those that can adapt, innovate, and capture the imagination of a new generation of consumers.

At its heart, entrepreneurship in craftsmanship is about creating value beyond the product. It's about building a brand that resonates with consumers on a personal level. This value creation extends beyond the economic; it's about contributing to the cultural and social fabric, preserving traditional skills while pushing the envelope of innovation.

The potential for growth and impact is immense. By harnessing the power of digital platforms, craftsmen can not only sustain their practices but also scale them, reaching diverse markets and opportunities previously beyond reach. This scalability is not just about financial growth; it's about the potential to influence and inspire, to set new standards in sustainability, craftsmanship, and entrepreneurial success.

For aspiring craftsman-entrepreneurs, the journey is both daunting and exhilarating. The keys to success include a commitment to lifelong learning, an openness to collaboration, and a willingness to embrace technology and innovation. With these tools, the modern craftsman can navigate the complexities of the digital age, turning challenges into opportunities for growth and innovation.

Through entrepreneurship and innovation, modern craftsmen are not just preserving their heritage; they are redefining it. They are building a future where craftsmanship is not just about what can be made with hands but what can be envisioned with the mind and realised with the heart. In doing so, they are charting a course not just for themselves but for future generations, illustrating that craftsmanship can be a dynamic, evolving practice that embodies the best of tradition and innovation.

As the landscape of craftsmanship continues to evolve, the fusion of entrepreneurship, innovation, and craft promises not just a renaissance of traditional trades, but the birth of new ones. In this new era, the craftsman's workshop is both a haven of creativity and a hub of innovation, where the legacy of the past and the promise of the future are crafted with equal skill.

In conclusion, the confluence of entrepreneurship and innovation within craftsmanship is more than a trend; it's a transformation. It's a testament to the resilience, creativity, and spirit of craftsmen who, in the face of changing times, choose to adapt, innovate, and thrive. For those embarking on this journey, it's a path paved with challenges but illuminated by the possibility of making a lasting impact on the world—one craft at a time.

Chapter 8:
The Culture of DIY: Generation Z Making Its Mark

In a world where digital technology and traditional craftsmanship intertwine, Generation Z stands at the forefront, heralding a new era in the do-it-yourself (DIY) culture. This generation, adept at navigating online platforms, has taken the ethos of DIY to remarkable heights, fostering communities that blur the lines between amateur and professional. The rise of online forums, video tutorials, and social media has democratised learning, enabling young creators to master skills, share insights, and collaborate on projects with a global reach. This chapter delves into how Generation Z is not just participating in but significantly shaping the maker movement, infusing it with fresh perspectives, sustainability concerns, and a relentless pursuit of innovation. Their engagement is characterised by an embracing of open-source resources, a keen interest in eco-friendly materials, and a determination to renew traditional practices through modern technology. As they carve their niches, these young craftsmen are not merely making things; they're making their mark, challenging us to redefine what it means to be an artisan in the 21st century. Through their stories, we see a vivid picture of a future where craftsmanship is inclusive, sustainable, and enriched by the digital age, a vision that beckons to all who value the blend of tradition and innovation.

Online Communities and Collaboration

In an age where digital transformation touches every aspect of our lives, the realm of craftsmanship and DIY culture has not been left behind. Generation Z, in particular, has harnessed the power of online communities and collaboration, melting the barriers of distance and time to redefine what it means to be a craftsman in the modern world.

The essence of these online communities lies not in the mere exchange of ideas but in the culture of collaboration they foster. Across forums, social media platforms, and dedicated DIY websites, young artisans find not just a repository of knowledge but a lively ecosystem willing to share, critique, and enhance each other's work. This dynamic interaction is fostering a new culture within craftsmanship, one that is vibrant, inclusive, and innovative.

For Gen Z, the Internet is not just a tool but a space of infinite possibilities. They leverage platforms like YouTube and Instagram not only to showcase their creations but to learn and teach skills across a vast range of disciplines. From woodworking to 3D printing, online tutorials and courses offer accessible knowledge that was once gatekept within traditional apprenticeship models.

Moreover, these digital platforms become spaces where support and encouragement are abundantly available. The journey of learning a craft can be daunting, filled with trials and errors. Online communities offer a cushion of support, where failure is seen as a step towards mastery, not a dead end.

Collaboration extends beyond emotional support and into real-world projects. Crowdsourcing ideas and feedback, young artisans can refine their designs and concepts through a wealth of perspectives. In some cases, collaborative projects transcend digital conversations into physical creations, with individuals from different parts of the world contributing parts to a singular piece, showcasing the incredible potential of global partnership.

This culture of online collaboration has also opened doors for digital craftsmanship. With technologies such as 3D printing and virtual reality, crafts are no longer confined to traditional materials. These digital artisans are exploring and creating art forms that were unimaginable a few decades ago, all facilitated through online collaboration and learning.

Equally important is the role of such communities in championing sustainable practices within craftsmanship. With a keen interest in addressing environmental issues, many online platforms dedicated to crafts promote the use of eco-friendly materials and upcycling techniques. Through tutorials, challenges, and shared projects, they're not only learning how to create but to do so responsibly, with an eye towards the future of our planet.

Online marketplaces and social media have also revolutionised how young craftsmen find their audience and market. By connecting creators directly with consumers, these platforms have opened up new possibilities for entrepreneurship in the crafts sector, allowing artisans to reach a global audience without the mediation of galleries or retail spaces.

In education, too, online platforms have had a transformative impact. With the rise of online courses and virtual workshops, learning a craft is no longer limited to those who can access physical trade schools or apprenticeships. This has democratised learning, making it possible for anyone with an internet connection to develop their skills and contribute to the culture of craftsmanship.

The heart of this digital transformation in craftsmanship lies in its community spirit. Online forums and networks serve as hubs for mentorship, where experienced artisans guide newcomers, offering advice, sharing resources, and sometimes even tools. This spirit of generosity and mutual growth has always been at the core of craftsmanship but is finding new expression in the digital age.

Despite the virtual nature of these interactions, the sense of community is palpable. Annual meet-ups, live streams, and collaborative projects help forge strong bonds among members, many of whom have never met in person. This creates a unique blend of the traditional sense of craftsmen guilds with the modern digital landscape.

However, navigating online communities requires a balance. The overwhelming amount of information and the fast-paced nature of social media can sometimes lead to burnout. Thus, many within the community actively promote discussions around mental health, stressing the importance of taking breaks and the value of offline crafting sessions.

It's clear that online communities and collaboration are at the forefront of redefining craftsmanship for Generation Z. By breaking down barriers, these digital platforms have not only made crafts more accessible but have also fostered a global community of young artisans eager to learn, share, and innovate together. As we move forward, the fusion of tradition and technology promises to bring forth a new era of craftsmanship, rich with creativity, collaboration, and sustainability.

In conclusion, the rise of online communities has imbued the world of craftsmanship with a new spirit. It is a spirit characterised by openness, diversity, and a shared commitment to growth and innovation. As this book explores the evolving landscape of craftsmanship, it becomes evident that these digital platforms are not just tools but catalysts for creating a vibrant, inclusive, and sustainable future for the trade. Through collaboration and community, Generation Z is indeed making its mark, crafting a legacy that bridges the past and future.

The Maker Movement

At the heart of Generation Z's engagement with crafting and small-scale manufacturing lies the Maker Movement, a cultural revolution that embraces technology, innovation, and traditional handicrafts in equal measure. This movement, heralding a new era of production, is not just about making things; it's about making things happen. It's where creativity meets functionality, and where the future of craftsmanship is being redefined.

The Maker Movement is founded on the principles of exploration, experimentation, and education. It embodies the spirit of curiosity that has fuelled advancements throughout history, but with a modern twist. Here, the garage inventor, the tech enthusiast, and the traditional craftsman walk the same path, albeit wielding different tools. It's a convergence of worlds, where 3D printers are as commonplace as woodworking tools.

Gone are the days when production was left solely to large manufacturers. The Maker Movement brings the power of creation back to the individual, championing a do-it-yourself ethic that Generation Z has embraced wholeheartedly. In this new paradigm, the barriers to entry for creating and innovating are lower than ever, thanks to accessible technologies and the sharing of knowledge across online communities.

One of the most significant impacts of the Maker Movement is its potential for personalisation. In a world increasingly filled with mass-produced goods, the desire for unique, custom-made items is growing. This movement offers a way to satisfy that desire, providing the tools and knowledge to create truly unique items that carry a personal touch.

Educationally, the Maker Movement has immense potential. It's not just about teaching hard skills like coding or electronics; it's about fostering a mindset of problem-solving and critical thinking. Schools and educators are beginning to recognise the value of hands-on, project-based learning that the movement champions, seeing it as a way to prepare students for the challenges of the future.

However, the Maker Movement isn't without its challenges. Accessibility remains an issue, with socioeconomic factors affecting one's ability to participate. Despite the lowering of technological barriers, economic ones still exist. Yet, there's hope. Initiatives aimed at sharing resources and knowledge are beginning to pop up, aiming to make the Maker Movement truly inclusive.

The role of online platforms cannot be overstated in the rise of the Maker Movement. Websites and forums offer not just tutorials and resources, but communities of supportive individuals eager to share their expertise and learn from each other. This collaborative spirit is a cornerstone of the movement, embodying the idea that we can achieve more together than we can apart.

Moreover, the Maker Movement is making its mark on the sustainability front. The movement encourages the use of eco-friendly materials and upcycling practices, aligning with the growing global concern for the environment. Here, innovation isn't just about creating new things; it's about reimagining the use of what we already have in sustainable ways.

Entrepreneurship is another area where the Maker Movement is having a profound impact. It's enabling individuals to turn their passions into professions, transforming hobbies into viable businesses. This entrepreneurial spirit is not just about financial gain; it's about the fulfilment that comes from doing what one loves and sharing it with the world.

The Maker Movement also serves as a bridge between generations. It's a space where the expertise of older generations intersects with the fresh perspectives of the younger ones, facilitating a transfer of knowledge that's mutually beneficial. This intergenerational exchange enriches the movement, ensuring the preservation of traditional skills while fostering innovation.

In essence, the Maker Movement is more than just a trend; it's a testament to the ingenuity and resilience of the human spirit. It shows us that with the right tools and a supportive community, there's almost no limit to what can be created. It empowers individuals to take control of their environment, to learn through doing, and to share their discoveries with the world.

Looking to the future, it's clear that the Maker Movement will continue to evolve. As new technologies emerge and societal needs change, so too will the ways in which we create and innovate. However, the core values of the movement - curiosity, creativity, collaboration, and community - will undoubtedly remain at its heart.

For educators, industry professionals, and young individuals alike, the Maker Movement offers a vision of what the future of craftsmanship could look like. It's a call to action to engage with the world in a more hands-on, sustainable, and meaningful way. In embracing the Maker Movement, we're not just crafting objects; we're crafting the future.

So let us embrace this movement with open arms and open minds. Let's explore, innovate, and create, not just for ourselves but for the generations to come. The Maker Movement is more than just a cultural phenomenon; it's a beacon of hope for a future where anyone can make their mark on the world, one creation at a time.

Chapter 9: Women in Trades: Breaking Stereotypes

In a world where the lens of tradition has often focused narrowly on gender roles, the narrative of women in trades stands as a beacon of progress and defiance. No longer confined to the peripheries of craftsmanship, women are not just entering trades historically dominated by men; they're redefining them, bringing a blend of skill, innovation, and perspective that enriches the field. This chapter delves into the heart of this transformation, exploring how the fusion of resilience and creativity has enabled women to carve out their niches in sectors from blacksmithing to digital fabrication. It's not merely about inclusion for its own sake; it's about how diverse perspectives drive innovation, how breaking the mould can lead to unforeseen successes and how women are not just participating in trades but are at the vanguard, leading them into the future. Their stories—of challenges faced, stereotypes shattered, and victories hard-won—don't just inspire; they serve as a compelling call to action for anyone, regardless of gender, envisioning a career built on craftsmanship, technology, and innovation. As we shine a light on these pioneers, it becomes clear that their journey is not solely about personal achievement but a collective stride towards a more inclusive, dynamic, and innovative industry.

Shifting Perceptions

The trades have long been dominated by a certain archetype, often excluding many who wish to contribute their skill and passion. This chapter discusses an evolving landscape where women are not only participating in but thriving within various trades, breaking stereotypes and reshaping societal perceptions. The shift is gradual but evident, as barriers are dismantled and inclusivity is championed.

In the past, certain trades were considered unsuitable for women, trapped in an outdated mindset that not only stifled women's aspirations but also deprived industries of diverse talents. However, the narrative is changing. One key driver of this change is the undeniable proof of capability and achievement women bring to these fields. From construction to carpentry, welding to masonry, women are proving that skill knows no gender.

Technology and innovation have been significant catalysts in this transformation. The modern craftsmanship landscape, powered by digital tools and technologies, demands a skill set that is not bounded by physical strength but rather creativity, precision, and a willingness to learn. This has levelled the playing field, allowing more women to enter and excel in trades that were once almost exclusively male.

Education and visibility play crucial roles in shifting perceptions. As more trade schools and apprenticeship programs actively recruit and support female students, young girls have visible role models to inspire them. The stories of successful women craftsmen, highlighted through various media, further serve to normalise women's presence in these fields, showing the world that they are equally capable of craftsmanship excellence.

Within the industry, there's an emerging culture of support and mentorship among women. Networks and associations specifically for women in trades provide not only professional guidance but also a sense of community. This camaraderie among female tradespeople is invaluable for the encouragement and empowerment it offers to both newcomers and veterans in the field.

However, challenges remain. Stereotyping and prejudice have not been entirely eradicated. Women in trades still frequently have to prove their worth beyond what is expected of their male counterparts. The battle against these outdated perceptions requires a collective effort from both within and outside the industry.

Employers play a crucial role in shifting perceptions. Those who actively implement policies that promote diversity and inclusivity set a precedent for others. By creating supportive working environments that encourage the participation of women, employers can significantly impact the demographic makeup of the trades.

Policy and advocacy work is also vital. Legislative changes that ensure equal opportunities and protect against discrimination are fundamental to creating an environment where women in trades can thrive. Advocacy groups continue to push for these changes, aiming for a future where one's gender is irrelevant to their career aspirations in trades.

The economic argument for diversity in trades is compelling. Studies have shown that diverse teams are more innovative and efficient, leading to better outcomes and profitability. Hence, encouraging women to enter these fields is not just a matter of equity but also a smart business strategy.

At the grassroots level, initiatives aimed at young girls such as workshops, coding camps, and mentorship programmes introduce them to the possibilities within trades. By catching their interest early, these initiatives help to sow the seeds for a more gender-diverse workforce in the future.

Societal attitudes towards women in trades are also influenced by media representation. The more we see women portrayed in these roles in films, television, and online platforms, the more normalized it becomes. This representation matters, as it shapes the perceptions of both current and future generations about what is possible and acceptable.

Innovative projects at the intersection of craftsmanship and modern technology often see significant contributions from women. These projects highlight the evolving nature of trades and the essential role of creativity and innovation, areas where women have excelled. This further helps to shift the narrative about who can be a craftsman in the digital age.

The journey towards gender inclusivity in trades is ongoing, and while there have been significant strides, there's still much ground to cover. The collective efforts of individuals, organisations, and governments are crucial in continuing this momentum. Empowering women to pursue their passions in trades not only enriches the industry with diverse talents but also challenges and ultimately changes the archaic stereotypes that have held back progress.

Encouragement from family and society remains a strong influence on young women considering careers in trades. As perceptions shift and the narrative of who can be a tradesperson evolves, it is hoped that future generations of women will face fewer barriers and broader acceptance in pursuing their craft.

In conclusion, the shifting perceptions about women in trades is a testament to the resilience, talent, and determination of countless women who have paved the way. It also signals a broader cultural shift towards recognizing and valuing diversity in all sectors. As we move forward, the integration of women into trades not only enriches the field but also ensures a more innovative, inclusive, and sustainable future for craftsmanship.

Success Stories

As we delve into the heartening realm of women's triumphs in the trades, it's crucial to understand that each story is not just about breaking stereotypes but is a beacon of inspiration, encouraging a new generation to redefine craftsmanship. This chapter encapsulates the journeys of women who have carved their niche in what is traditionally considered a man's world, illustrating the potential for innovation, resilience, and success against the odds.

The first narrative introduces Sarah, a welder who has not only mastered her craft but has also paved the way for sustainability in welding practices. Sarah's workshop in the outskirts of Birmingham utilises solar power for operations, showcasing her commitment to eco-conscious craftsmanship. Her work, predominantly in sculptural installations, has been displayed in galleries across Europe, proving that the integration of traditional skills with a modern approach can lead to remarkable achievements.

Emma's story brings to light the transformative potential of technology in crafts. A carpenter by trade, Emma has harnessed the power of 3D printing to create bespoke furniture pieces. Her innovative approach has not only reduced material waste but has also allowed for intricate designs that would be impossible to achieve by hand alone. Emma's workshop in Bristol has become a hub for aspiring carpenters keen on exploring the intersection of technology and traditional woodworking.

Journeying north to Glasgow, we encounter Rachel, an electrician who is revolutionising her field with smart technology. Rachel's installations in residential homes focus on enhancing energy efficiency and integrating IoT devices, aligning with the growing demand for smart homes. Her success lies in her ability to combine her trade skills with a deep understanding of digital innovations, setting a new standard for electricians in the 21st century.

In Sheffield, Liz, a metalworker, has turned her passion for recycling into a flourishing business. By repurposing discarded metals into artistic and functional items, Liz not only contributes to reducing waste but also brings attention to the importance of sustainability in craftsmanship. Her creations range from bespoke jewellery to unique home decor, each piece telling a story of transformation and resilience.

Not to be overlooked is Anna, a stonemason from Cornwall, whose precision and artistic flair have restored numerous historical buildings, preserving the heritage of craftsmanship for future generations. Anna's dedication reveals the potential of trades to contribute not only to individual success but also to community and cultural heritage.

These women's stories are complemented by Jane, a glassblower whose studio in Kent is pioneering the use of renewable energy sources in glass production. Her work, characterised by vibrant colours and sustainable practices, challenges the perception of glassblowing as an energy-intensive craft.

Moreover, Sophie, a bespoke tailor in London, is redefining the traditional world of suit-making. By integrating smart fabrics that adapt to the wearer's body temperature and motion, Sophie's creations are at the forefront of wearable technology, marrying the elegance of tailoring with the innovations of the digital age.

Moving across the Atlantic, we find Mia in New York, a ceramist whose pottery integrates solar cells, turning everyday objects into power sources. Mia's approach not only revives the ancient craft of pottery but also repurposes it for contemporary environmental challenges, illustrating the global impact of innovative craftsmanship.

Back in the UK, in Leeds, Patricia, a landscape gardener, employs augmented reality to design gardens that are not only beautiful but also biodiverse havens. Her work emphasises the crucial role of trades in addressing climate change, showcasing how traditional skills, combined with new technologies, can contribute to creating sustainable environments.

Lastly, we introduce Erin, a software developer turned jeweller, whose pieces incorporate QR codes that narrate the story of their creation. Based in Manchester, Erin's journey from tech to crafts highlights the seamless blend of digital and physical production, offering a new narrative for the future of artisanry.

These stories embody the spirit of innovation, resilience, and mastery that defines the modern craftsman. They are not just success stories; they are invitations to rethink what is possible in the trades. Women across the globe are not just participating in craftsmanship; they are leading it, transforming it, and in doing so, they are reshaping our world.

Each narrative underscores the importance of perseverance, curiosity, and the willingness to challenge traditional boundaries. The women featured in this chapter have not only forged successful careers but have also contributed to the broader dialogue on sustainability, technology, and inclusivity in the trades.

As we reflect on these stories, it's clear that the fusion of traditional skills with modern technologies and sustainable practices is not just the future of craftsmanship; it's a vibrant present. Women in trades are not just breaking stereotypes; they are redefining what it means to be a craftsperson in the 21st century, inspiring generations to come to embrace innovation, respect tradition, and pursue their passions with determination.

In conclusion, the success stories of women in trades serve as powerful testaments to the evolving landscape of craftsmanship. They highlight the transformative power of integrating innovation with tradition, the importance of sustainability, and the limitless possibilities that emerge when barriers are broken. These narratives are not just inspiring; they are a call to action, urging us all to support, celebrate, and participate in the vibrant future of trades.

Chapter 10: The Role of Social Media in Modern Craftsmanship

In today's digital age, social media has transcended its original purpose of simple communication, developing into an indispensable tool for modern craftsmen. It empowers artisans with the ability to not just showcase their unique creations to a global audience but also to tell their stories, engage with followers, and build communities centered around shared interests and values. This paradigm shift has opened new avenues for branding and marketing, allowing craftsmen to connect directly with their clientele and fellow artisans without the need for intermediaries. Platforms like Instagram, Pinterest, and Etsy have become digital galleries, where the visual allure of handmade goods can be displayed in its full splendour, captivating the hearts and minds of consumers who value authenticity, sustainability, and the sheer beauty of handcrafted items. Furthermore, social media serves as a vibrant forum for exchanging ideas, learning new techniques, and fostering collaboration, thereby nurturing a culture of innovation and continuous improvement. For the modern craftsman, mastering social media is not just about keeping up with the times; it's about embracing the future of their trade, ensuring their art thrives and is appreciated across generations and geographical boundaries. As we delve deeper into this topic, we'll explore strategies for effectively branding and marketing crafts, as well as the importance of community building and how these elements contribute to the sustainability and growth of contemporary craftsmanship.

Branding and Marketing Your Craft

In today's interconnected world, the art of crafting isn't just limited to the physical act of creation. A substantial aspect of modern craftsmanship involves how one presents and promotes their work to a wider audience. This chapter delves into the essential strategies for branding and marketing your craft, leveraging the digital landscape to elevate your work from a local art form to a globally recognised brand.

Branding, in essence, is about storytelling. It's the narrative you craft around your work that invites your audience into your world. The most compelling brands in craftsmanship go beyond just showcasing their creations—they share the journey of their making, the passion behind the hands that craft, and the ethos that drives their work. It's about constructing a persona for your craft that resonates with your audience on a personal level.

Marketing, while often intertwined with branding, focuses more on the strategies and platforms used to communicate your brand's story to the world. In the age of social media, these platforms become powerful tools in the craftsman's toolkit. Sites like Instagram, Pinterest, and Etsy are not just channels for exposure; they are communities where interaction and engagement can transform casual viewers into loyal customers and advocates for your brand.

But how does one begin? Identifying your unique value proposition is the first step. What makes your craftsmanship stand out? Is it the traditional techniques you employ, the innovative use of materials, or perhaps the sustainability of your practice? Pinpointing this helps in tailoring your brand narrative to highlight what sets you apart in the vast sea of global artisans.

Visual storytelling becomes indispensable in branding and marketing your craft. High-quality photographs and videos that capture not just the finished product but the process of creation add depth to your narrative. They invite your audience to appreciate the meticulous effort and creativity that goes into each piece. Remember, in the digital realm, your visuals are often the first point of contact with potential customers. Make them count.

Engaging consistently with your audience is another cornerstone of effective marketing. Respond to comments, share behind-the-scenes glimpses, and perhaps involve your followers in some of the decision-making processes. Modern consumers value authenticity and connection; they want to support brands with whom they feel a personal bond.

Email marketing, though seemingly traditional, remains a potent tool. By inviting visitors to your site to subscribe to your newsletter, you create a direct channel for communication. This can be invaluable for announcing new collections, sharing exclusive content, or offering special promotions.

Collaborations with other craftsmen or brands can also serve as a dynamic marketing strategy, introducing your work to an entirely new audience. These partnerships, whether for a single project or an ongoing series, can generate buzz and foster a sense of community within your niche.

SEO (Search Engine Optimization) is a term that might seem daunting but is crucial in ensuring your website or online store ranks high in search results. Basic practices like using the right keywords in your product descriptions and blog posts, optimizing images, and earning backlinks from reputable sites can significantly increase your online visibility.

Attending and participating in craft fairs and exhibitions is another vital strategy. These events provide an opportunity not just for selling your products but for networking, learning from peers, and gaining direct feedback from customers. They also serve as a physical extension of your brand, allowing you to create a real-world experience that reflects your online presence.

Storytelling extends to packaging as well. Your product's packaging is the final touchpoint in the customer experience and should reflect the care, quality, and values of your craft. Sustainable, innovative, or beautifully designed packaging can make unboxing a memorable part of the buyer's journey, encouraging them to share their experience with others.

Influencer marketing, where craftsmen team up with social media influencers to reach a larger audience, can also be incredibly effective. The key is to find influencers whose values align with your brand, ensuring their endorsement feels authentic to their followers.

Analytics should not be overlooked in your marketing strategy. Tools like Google Analytics or insights from social media platforms provide valuable data on who your audience is, what content they engage with the most, and which marketing channels are most effective. This information is crucial for refining your strategies and ensuring that your efforts yield the best results.

Patiently cultivating your brand and marketing your craft is a journey, not a race. It's about consistent effort, learning from feedback, and adapting to the ever-evolving digital landscape. Above all, it's about staying true to the essence of your craft while finding innovative ways to share it with the world.

As we move forward, embracing the tools and platforms available in the digital age, remember that at the heart of your brand is your unique craft. The story, quality, and passion behind your work are what will ultimately resonate with people. Branding and marketing are merely the channels through which your craft's message is amplified. Grounded in authenticity and driven by innovation, your brand has the potential not just to reach but to touch and inspire a global audience.

Building a Community

In the modern era, where individualism often takes centre stage and digital platforms dominate our communication methods, the art of building a community around craftmanship has evolved significantly. Social media, with its boundless reach and potential for real-time interaction, has become a linchpin in this transformation. Its role in fostering a sense of belonging and shared purpose amongst craftsmen and enthusiasts cannot be understated.

At the heart of every thriving craft community on social media is the storytelling prowess of its members. These platforms allow artisans to share not just the final outcome of their labours but the process, passion, and perseverance that go into each work. By pulling back the curtain on their creative process, craftsmen connect with their audience on a more personal level, inviting them into their world.

The shared narratives of trial and error, success, and sometimes failure, resonate deeply with followers, who see in these stories the universal truths of hard work and resilience. Thus, a community begins to grow, bound not by geography, but by shared values and mutual respect for the craft.

Interactive features of social media platforms, such as comments, likes, and shares, further empower this community building. They provide an avenue for immediate feedback, encouragement, and constructive criticism, facilitating a dynamic conversation between craftsmen and their audience. This two-way communication is foundational in creating an engaged and active community.

Moreover, social media offers a unique platform for collaboration among craftsmen. Across continents and time zones, artisans can work together on projects, exchange ideas, and learn from each other. These collaborations, often documented and shared on social media, inspire further interaction and community growth.

Education plays a crucial role within these communities as well. Experienced craftsmen pass on their knowledge to novices, not through formal apprenticeships, but via tutorials, live sessions, and Q&As directly on social media. This informal mentorship enriches the community, ensuring the transfer of skills and traditions to new generations.

The inclusivity afforded by social media stands out as well. Irrespective of one's background, location, or level of expertise, all are welcome to join these craft communities. This openness contributes to a diverse and vibrant community, an essential aspect of innovation and creativity.

Events and meet-ups organised through social media also play a critical role in strengthening craft communities. Though much interaction happens online, the importance of physical gatherings where members can meet, share their works, and engage in workshops cannot be overstated. These events bridge the virtual gap, creating strong bonds amongst members.

Social media additionally provides a platform for craftsmen to tackle challenges together. Whether it's addressing environmental concerns, sourcing sustainable materials, or advocating for fair trade practices, these communities have the collective power to effect change. Through campaigns and initiatives, they can raise awareness and push for action both within and beyond their communities.

The support system inherent in these communities is particularly beneficial for independent craftsmen and small businesses. Through shared promotion, featuring each other's work, and providing moral support during challenging times, members help each other to grow and thrive.

Feedback loops created within these communities are invaluable. They allow craftsmen to refine their work based on the collective input of their community, pushing them to innovate and improve continually. This iterative process is accelerated by the immediate interactions social media facilitates.

The beauty of building a community in the digital age lies in its global reach. Craftsmen are no longer confined to local markets or networks. Through social media, they can connect with like-minded individuals across the globe, opening up unprecedented opportunities for collaboration, learning, and growth.

Despite the vast and sometimes overwhelming nature of social media, niche communities have flourished. These smaller, focused groups allow for deeper connections and more targeted discussions, fostering a strong sense of identity and belonging within the larger craft community.

In conclusion, the role of social media in building a community around craftsmanship is profound. It breaks down barriers, connects disparate individuals, and fosters a shared identity and purpose. Through storytelling, collaboration, education, and support, these digital communities not only enrich the lives of their members but also play a crucial role in sustaining and evolving the rich tapestry of modern craftsmanship.

In this era, where technology and tradition intertwine, social media stands as a testament to the human need for connection and community. As craftsmen continue to navigate the digital landscape, building and nurturing these communities will remain pivotal in ensuring the vibrancy and resilience of craftsmanship in the modern age.

Chapter 11: Balancing Tradition and Innovation

In the heart of modern craftsmanship lies a crucial duel between preserving time-honoured traditions and embracing groundbreaking innovation. This chapter elucidates how today's craftsmen can hold onto the essence of their trade's history while also propelling it into the future with cutting-edge technologies and methodologies. The challenge of keeping traditions alive is not about resisting change but about integrating new ideas in ways that respect and revitalise those traditions. Innovative approaches to craftsmanship are not just about adopting new tools or techniques for the sake of novelty; they're about enhancing efficiency, accessibility, and sustainability, thereby ensuring the relevance of trades in today's fast-paced world. As we delve deeper, we find that the symbiosis of tradition and innovation offers an invigorating path forward, one that can enrich the craft, expand its appeal, and ensure its survival for future generations. The key lies in recognising that innovation isn't antithetical to tradition but is, in fact, its greatest ally. By approaching craftsmanship with a mindset open to both honouring the past and embracing the future, craftsmen can create works that resonate with a broad audience, encapsulating the beauty and durability of time-tested methods while pushing the boundaries of what's possible in the realm of the handmade.

The Challenge of Keeping Traditions Alive

In the tapestry of human history, the threads of tradition and innovation are interwoven, creating a rich cultural fabric that defines societies. Yet, as we venture further into the 21st century, we find ourselves at a crossroads where the preservation of these very traditions seems increasingly challenging. The quandary of maintaining the essence of traditional craftsmanship while embracing the inevitable tide of technological advancement is not merely an academic concern but a practical reality faced by artisans worldwide.

Traditional crafts, honed over centuries, serve as a tangible connection to our past. They encapsulate the wisdom, skills, and aesthetics of generations past, shaping our understanding of cultural identity and heritage. However, the rapid pace of technological evolution and globalisation poses a significant threat to these age-old practices. The commodification of mass-produced goods, coupled with the dwindling transmission of artisanal knowledge from one generation to the next, casts a shadow over the future of traditional crafts.

The challenge lies not in halting the march of progress but in harmonising the old with the new. It is about finding a middle ground where innovation enhances rather than erodes the value and charm of traditional craftsmanship. This delicate balance requires a conscientious effort from all stakeholders involved, including artisans, educators, industry experts, and especially the younger generations who stand to inherit this legacy.

Engaging Millennials and Gen Z, known for their affinity with technology and innovation, in the world of traditional crafts presents both an opportunity and a challenge. These generations value sustainability, authenticity, and personalisation – qualities inherent in traditional craftsmanship. By aligning the intrinsic values of traditional crafts with the aspirations of younger generations, we can rekindle interest and ensure the transfer of valuable skills and knowledge.

However, the reimagination of apprenticeships and education in craftsmanship is crucial. Traditional pedagogical approaches may not resonate with the digital-native generations. Instead, integrating technology, such as online learning platforms and virtual reality, into the learning process can make the acquisition of these age-old skills more appealing and accessible.

The role of social media and online communities cannot be overstated in this endeavour. By leveraging these platforms, artisans can showcase their crafts, share their stories, and connect with a global audience. This not only aids in the preservation and promotion of traditional crafts but also helps in building a supportive community of young enthusiasts eager to learn and participate.

Furthermore, the incorporation of sustainable practices into traditional crafts aligns with the growing global consciousness regarding environmental sustainability. It offers a compelling narrative that appeals to younger generations, who are increasingly aware of their ecological footprint. Sustainable craftsmanship, therefore, represents a confluence of tradition and contemporary values, making it a vital area for innovation.

Innovation in materials and techniques also plays a pivotal role in keeping traditional crafts alive. Experimenting with eco-friendly materials, employing state-of-the-art tools to achieve precision, or utilizing digital platforms for design and collaboration can enhance the appeal and functionality of traditional crafts, ensuring they meet the demands and expectations of modern consumers.

Yet, the adaptation of new technologies and practices must be sensitive to the cultural significance of crafts. The essence of traditional craftsmanship lies in its connection to history and heritage. Therefore, innovation should not dilute this essence but rather preserve and accentuate it. Educators and industry professionals must guide this integration, ensuring that technological advancements serve as tools for preservation rather than agents of change.

The economic dimension of craftsmanship in the digital age also necessitates a nuanced understanding. As the marketplace becomes increasingly global, traditional artisans face the challenge of distinguishing their wares in a saturated market. Herein lies the opportunity for entrepreneurship and innovation by crafting unique brand stories, leveraging online marketing, and targeting niche markets interested in authentic, handcrafted products.

Women in trades and the breaking of stereotypes also form a crucial part of revitalizing traditional crafts. Encouraging and supporting the participation of women not only aids in the preservation of craftsmanship traditions but also fosters diversity and innovation within the craft community.

Cross-cultural exchanges and collaborations further enhance the sustainability and relevance of traditional crafts. Such interactions can inspire new fusion techniques, styles, and materials, enriching the craft heritage and fostering greater appreciation and understanding across cultural boundaries.

In conclusion, the challenge of keeping traditions alive in an era of unprecedented technological and social change is daunting but not insurmountable. It requires a collaborative effort, a fusion of old and new, and a deep reverence for the craftsmanship that connects us to our past. By embracing innovation judiciously, fostering education and community, and promoting sustainability and inclusion, we can ensure that traditional crafts not only survive but thrive in the modern era.

The path forward is not one of resistance to change but one of adaptation and evolution. As we forge ahead, let us carry the torch of our ancestors, illuminating the future with the brilliance of traditional craftsmanship, redefined for a new generation. In this journey, we all have a role to play: as artisans, as educators, as consumers, and as custodians of our cultural heritage. The challenge of keeping traditions alive is ours to meet, with creativity, passion, and perseverance.

Innovative Approaches to Craftsmanship

In an age where tradition often collides with innovation, there exists a vibrant landscape for craftsmen where the amalgamation of time-honoured techniques and modern technology not only coexists but flourishes. The craftsmen of today are standing on the shoulders of giants, leveraging centuries of accumulated knowledge to forge new paths and redefine what it means to be a master of their trade.

The narrative of craftsmanship has always been one of evolution. From the earliest tools and artefacts to the refined products of today, each generation of craftsmen has contributed its chapter to this ongoing saga. However, the introduction of digital technologies in the 21st century has accelerated this evolution, bringing forth a paradigm shift in how we perceive and interact with the act of making.

At the heart of this transformation lies the digital artisan – a new breed of craftsmen who harnesses the capabilities of digital fabrication technologies, such as 3D printing and laser cutting, to expand the bounds of what can be created. These technologies, once the exclusive domain of industrial production, have now been democratized, enabling artisans to prototype and iterate their designs with unprecedented speed and precision.

But innovation in craftsmanship isn't just about embracing new technologies; it's also about rethinking traditional materials and methodologies. The use of sustainable, eco-friendly materials is gaining traction among artisans who are keen to reduce their environmental footprint. Methods such as upcycling and recycling are not just seen as ethical choices but as sources of inspiration and creativity.

Education and skill development have also evolved to keep pace with these changes. Traditional apprenticeships are being reimagined to include digital competencies, while online platforms and virtual reality tools are being used to train and inspire the next generation of craftsmen. This blend of hands-on experience with digital education is forging a more adaptable and versatile workforce.

Another dimension of innovation in craftsmanship is the resurgence of the DIY culture, fueled in part by online communities and collaborative platforms. The maker movement, born out of a desire to return to the basics of making and a dissatisfaction with mass-produced goods, highlights a growing appreciation for the unique and the bespoke.

The role of social media and branding in modern craftsmanship cannot be overstated. Artisans today have at their disposal powerful tools to market their crafts, tell their stories, and build communities. This has not only opened up new avenues for customer engagement but has also enabled craftsmen to achieve a global reach that was previously unimaginable.

Yet, with all these advancements, the challenge remains to keep traditions alive. It's a delicate balance between preserving the essence of traditional craftsmanship and embracing the possibilities offered by innovation. Some purists fear that technology may dilute the authenticity of handcrafted products, yet others argue that innovation can enhance rather than eclipse traditional methods.

This brings us to an interesting juncture — where the defining factor of craftsmanship is no longer about choosing between traditional methods and modern technology, but about how seamlessly one can integrate the two. It's about leveraging technology to refine and expand traditional techniques, and vice versa.

Furthermore, the pursuit of innovative craftsmanship is not just about the end product but also about the process. It's about embedding sustainability, fostering inclusivity, and ensuring that the tradition of making remains a vibrant and dynamic force in society. Modern craftsmen are not just artisans; they are educators, innovators, and custodians of culture.

The economic landscape for craftsmen in the digital age also presents new opportunities and challenges. The rise of online marketplaces and the gig economy has facilitated new models of entrepreneurship and innovation. Craftsmen can now operate in a more agile and flexible manner, reaching out to niche markets and leveraging technology to streamline their operations.

Despite these advances, the future of craftsmanship hinges on our ability to maintain a sense of community and shared purpose among craftsmen. As we navigate the complexities of the digital age, it's essential that we foster environments where knowledge can be shared, traditions can be honoured, and innovation can be celebrated.

As we look towards the future, it's clear that the path of craftsmanship will be one of convergence, where the rich tapestry of traditional techniques is interwoven with the threads of innovation and technology. It's a future where craftsmanship is not just about preserving the past but about imagining and crafting the future.

In this journey of transformation, education, mentorship, and community engagement will play pivotal roles. By empowering the next generation of craftsmen with the skills, knowledge, and mindset to navigate this evolving landscape, we can ensure that the art of making remains vibrant and relevant for generations to come.

In conclusion, the innovative approaches to craftsmanship remind us that at the heart of every crafted piece lies a story of human ingenuity and creativity. As craftsmen continue to navigate the delicate balance between tradition and innovation, they not only honour the legacy of their forebears but also lay the foundations for a future where craftsmanship continues to inspire, innovate, and thrive.

Chapter 12: Mental Health and Well-being in the Trades

In delving into the essence of mental health and well-being within the trades, we must understand that the craftsmanship journey compels more than just a physical commitment; it is an intimate dance of the mind and spirit. The act of creating, with one's hands intertwined with the threads of imagination and reality, offers a therapeutic solace unmatched by conventional means. It's here, in the rhythmic hum of the workshop, amidst shavings of wood or the soft glow of molten metal, that many find a meditative peace, an escape from the relentless pace of the digital age. The importance of fostering robust support networks and communities becomes ever more evident, serving not only as a bastion for sharing knowledge and skills but also as a critical lifeline for mental health resilience. As we navigate the intricacies of this chapter, we are reminded that the well-being of craftsmen is as much about the environment they are part of as it is about the individual. Through collective endeavour, the crafting community can offer a beacon of support, championing not only the preservation and innovation of the trades but nurturing the mental health of its artisans. In doing so, we pave a path not only towards a future enriched by skilled hands but also towards a well-being that elevates the soul of the craftsman in the pursuit of their art.

The Therapeutic Aspect of Crafting

In today's fast-paced world, where screens often dominate our attention and high-stress levels are all too common, crafting emerges not just as a hobby, but as a profoundly therapeutic undertaking. The act of creating something with one's hands is more than an escape; it's a return to the self, an intimate dance between the mind and the material world. This chapter delves into how crafting, a seemingly simple act, can wield immense power over our mental health and overall well-being.

The notion that manual work could serve as a form of therapy is far from new. History is replete with instances where the rhythmic motions of knitting, the precision of woodworking, or the focus required for intricate embroidery provided solace and a sense of accomplishment. These activities, deeply embedded in the fabric of traditional trades, carry with them an inherent therapeutic quality that remains relevant in the digital age.

One might wonder, in an era defined by technological advancement, why turn to age-old practices for mental wellness? The answer lies in the tactile experience of crafting. Engaging with physical materials offers a grounding effect, a much-needed anchor in a world that often feels ethereal and disconnected. The tangible nature of crafts enables individuals to connect with the here and now, fostering mindfulness and reducing stress levels.

The repetitive actions inherent in many crafts, such as the steady stitch of a needle or the methodical chiselling of wood, can induce a meditative state. This state of flow, a concept popularised in positive psychology, is where individuals become so immersed in an activity that all else fades away. In this space, time loses its grip, and the mind can rest from the constant barrage of thoughts and worries that characterise modern life.

Apart from offering a meditative reprieve, crafting can bolster self-esteem and confidence. Completing a project, whether it's a ceramic vase or a leather wallet, provides a tangible sense of achievement. In an age where accomplishments are often measured by likes and views, the satisfaction derived from crafting something tangible, something that can be held, used, or gifted, brings a unique and deeply personal fulfilment.

Moreover, crafting can act as a conduit for emotional expression. The colours, textures, and patterns one chooses can reflect internal landscapes, offering a non-verbal mode of expression for feelings that are hard to articulate. For many, the act of creating becomes a form of therapy in itself, a safe space to process emotions and achieve a state of emotional balance.

On a social level, crafting has the power to connect. Community workshops, crafting circles, and online forums offer spaces where individuals can share their work, seek advice, and find camaraderie. These communities provide not only technical support but also a sense of belonging, reducing feelings of isolation and loneliness.

For those struggling with anxiety or depression, the structure provided by crafting can offer a lifeline. The need to focus on a task can help redirect attention from negative thoughts, providing a break from the cycle of rumination that often accompanies these conditions. Crafting's inherent problem-solving – figuring out a pattern, correcting mistakes – can also enhance cognitive flexibility, a valuable skill in managing mental health.

The educational aspect of crafting can't be overlooked. Learning a new skill is not only rewarding but also stimulates the brain, offering cognitive benefits such as improved memory and concentration. For many, especially the elderly, crafting can be a vital tool in maintaining cognitive function and staving off the decline associated with aging.

Importantly, the therapeutic benefits of crafting are accessible. One doesn't need to be highly skilled or possess expensive materials to start. The simplicity of starting a project with just a few basic supplies makes crafting an inclusive activity, bridging generational and socioeconomic gaps.

Crafting also offers an antidote to the digital overload many of us face. In turning off screens and focusing on a tangible task, individuals find a much-needed break from the constant connectivity that characterises modern life. This disconnection allows the mind to recharge, improving attention spans and reducing digital fatigue.

In the trades, where the mastery of skill and creation of physical objects are core to the profession, the therapeutic aspect of crafting takes on additional significance. It reminds us that these trades are not just about the end product but about the process – a mindful, restorative journey that benefits both creator and community.

As we look to the future of craftsmanship, embracing its therapeutic aspects could be key. In a world that values efficiency and productivity, crafting offers a counter-narrative – one that champions slow, deliberate creation and the mental health benefits that come with it.

Integrating crafting into educational curriculums, workplace wellness programs, and community initiatives could provide an essential outlet for creativity and stress relief. By valuing the process over the product, we can foster environments that promote mental well-being, creativity, and a deeper connection to the world around us.

In conclusion, the therapeutic aspect of crafting is a testament to the timeless value of working with one's hands. It's a reminder that, in every stitch, cut, or mould, there lies the potential not just to create but to heal. As we navigate the complexities of modern life, crafting emerges not just as a skill, but as a salve, offering a path to a more balanced, mindful, and connected existence.

Community and Support Networks

In the intricate tapestry of mental health and well-being within the trades, the threads of community and support networks are both vibrant and vital. They serve not just as safety nets but as catalysts for growth, innovation, and resilience among craftsmen. The journey of every artisan, whether drenched in tradition or pioneering at the frontier of innovation, is undeniably challenging. It's a path beset with obstacles that are not just physical but deeply emotional and mental as well.

Understanding the profound impact of these networks, it's essential to delve into how they function and why they're so crucial. At their core, communities within the trades act as echo chambers of shared experiences, wisdom, and support. They provide a platform where voices, often muffled by the din of machinery or the solitude of meticulous labour, can be heard and acknowledged.

The advent of digital technology has significantly amplified the power and reach of these networks. Online forums, social media groups, and virtual meet-ups have broken down geographical barriers, enabling artisans from across the globe to connect, share, and learn from each other. This digital transformation has made support more accessible, immediate, and diverse.

Yet, the essence of these networks lies not just in the sharing of technical knowledge or market trends but in the mutual emotional support they provide. Crafting, by its nature, can be a solitary endeavour, with many artisans spending hours on end in their workshops or studios. In such environments, the feeling of isolation can be overwhelming. Here, community networks serve as a lifeline, offering companionship, empathy, and motivation.

Moreover, the role of mentorship within these networks cannot be overstated. Experienced craftsmen often take on the mantle of mentors, guiding newcomers through not just the technical aspects of their trade but also helping them navigate the mental and emotional challenges that come with the territory. This mentorship, steeped in a tradition of passing down knowledge from one generation to the next, is a cornerstone of the crafting community's support system.

Another significant aspect of these networks is their role in promoting mental wellness and resilience. Workshops, webinars, and talks focusing on mental health, stress management, and work-life balance are becoming increasingly common. They play a crucial role in de-stigmatizing mental health issues within the community, encouraging individuals to seek help and share their struggles without fear of judgment.

The inclusive nature of these communities also fosters a sense of belonging among members. It doesn't matter if you're a novice or a seasoned professional, your voice and experience are valued. This inclusivity is especially important in an era where the trades are seeing more diversity in terms of gender, ethnicity, and socio-economic backgrounds.

It's also worth noting how these networks catalyse innovation. By facilitating the exchange of ideas and collaborative projects, they create an environment where creativity flourishes. Members are encouraged to think outside the box, challenge norms, and explore new techniques and materials.

However, building and nurturing these support networks isn't without its challenges. It requires effort, trust, and a shared commitment to the collective well-being of its members. Leaders within these communities play a pivotal role in setting the tone, fostering a culture of openness, and ensuring that the networks remain responsive to the members' evolving needs.

Success stories abound, where individuals have overcome substantial hurdles through the strength drawn from their community. Whether it's navigating the complexities of turning a passion into a business, coping with the physical toll of the craft, or battling mental health issues, the support network often emerges as a critical factor in their triumph.

Yet, for all their benefits, the importance of these networks remains underappreciated in broader societal and policy contexts. Recognising and investing in support networks as essential infrastructures can propel the trades forward, ensuring they remain vibrant, sustainable and inclusive.

As the landscape of craftsmanship continues to evolve, so too must the nature and function of these support networks. They must adapt to new technologies, changing market dynamics, and the shifting needs of their members. But one thing remains constant - their indispensability. Without them, the craftsmanship journey would be a much lonelier and daunting endeavour.

Therefore, it's imperative for all stakeholders - from craftsmen and trade organisations to policymakers and educational institutions - to acknowledge, support, and invest in the development and sustainability of these networks. Their role in ensuring the mental health and well-being of craftsmen, fostering innovation, and securing the future of the trades cannot be overstated.

In conclusion, the community and support networks within the trades are not just a backdrop but the very fabric that holds the craftsmanship ecosystem together. They embody the spirit of collective resilience, innovation, and solidarity that is essential for navigating the challenges of the modern world. As we look towards the future, nurturing these networks should be at the heart of our efforts to support the next generation of craftsmen, ensuring that they not only survive but thrive in this ever-changing landscape.

Chapter 13: The Influence of Global Cultures on Craftsmanship

In an increasingly interconnected world, the cross-pollination of global cultures has woven a rich tapestry of inspiration for modern craftsmen and women. Chapter 12 delves into how these international influences have not only enriched traditional trades but also propelled them into the future with a new sense of diversity and vibrancy. Far from diluting traditional craftsmanship, the fusion of techniques and styles from different cultures has injected a fresh lease of life into time-honoured practices, enabling craftsmen to innovate while paying homage to the past. This cultural amalgamation encourages a deeper understanding and appreciation for the myriad ways in which different societies approach the art of making. Whether it's the meticulous precision of Japanese joinery influencing Western furniture-making or the vibrant patterns of African textiles being reimagined in contemporary fashion, the impact is profound. This exchange of ideas doesn't just revitalise traditional crafts; it fosters a global community of makers united by a shared passion for creation. In recognising the importance of these cross-cultural exchanges, we're reminded that the future of craftsmanship isn't confined within the borders of any one country. Instead, it thrives on the free flow of ideas, techniques, and inspirations that crisscross the globe, ensuring that the crafts of tomorrow are as diverse and dynamic as the world we live in.

Cross-Cultural Exchanges

In the fabric of today's global society, the thread of cross-cultural exchanges weaves a rich tapestry, uniting distant corners of the world through the shared language of craftsmanship. It's in the mingling of techniques, styles, and traditions that we witness a vibrant evolution of trades, one that promises to enrich the practice of craftsmen around the globe.

The impulse to integrate disparate cultural elements into craftsmanship is not merely a by-product of globalisation but a testament to humankind's inherent desire to connect, learn, and grow. As we delve into the cross-cultural exchanges shaping the present and future of trades, it becomes clear that these interactions are not just transforming craftsmanship; they're redefining it.

Consider, for example, the infusion of Japanese woodworking techniques into Western furniture making. This blend has given rise to pieces that not only showcase the minimalist elegance characteristic of Japanese aesthetics but also embody the robustness and functionality esteemed in Western design. Such exchanges challenge and expand our understanding of what is possible within the realm of craftsmanship.

Furthermore, the digital age has facilitated an unprecedented flow of knowledge across borders, enabling artisans from one corner of the world to access and draw inspiration from the craftsmanship traditions of another. Online platforms and social media have become conduits for this knowledge, allowing techniques that once might have taken generations to traverse continents to spread in mere seconds.

This rapid exchange, however, also poses challenges. The delicate balance between preserving the integrity of traditional crafts and embracing innovation is a tightrope that craftsmen must walk with care. It's imperative that in our enthusiasm for cross-cultural fusion, we don't dilute the essence that makes each tradition unique.

Moreover, the respect for the origins of these crafts is paramount. As craftsmen incorporate elements from various cultures into their work, acknowledging the roots and meaning of these influences fosters a deeper appreciation and prevents the superficial appropriation of surface-level aesthetics.

The benefits of cross-cultural exchanges in craftsmanship extend beyond the enrichment of the craft itself. They act as bridges of understanding among cultures, breaking down stereotypes and fostering mutual respect. Through the lens of shared crafts, we're reminded of our common humanity and the universal language of creativity.

The education sector, particularly in the field of arts and trades, has a significant role to play in nurturing this understanding. Curriculum that highlights the global dimensions of craftsmanship and encourages students to explore and integrate diverse cultural techniques can cultivate a new generation of artisans who are not only skilled but culturally sensitive and innovative.

Trade fairs and exhibitions also provide crucial platforms for cross-cultural engagement. By showcasing their work on a global stage, craftsmen have the opportunity to exchange ideas, collaborate on projects, and broaden their perspectives. These gatherings, in essence, become melting pots of cultural exchange, pushing the boundaries of traditional crafts and paving the way for novel creations.

Innovation, a key driver of modern craftsmanship, thrives on such diversity. The fusion of techniques and ideas from different cultures sparks creativity, leading to breakthroughs that might have remained undiscovered within the confines of a single cultural context.

The sustainability movement within craftsmanship is another area ripe for cross-cultural learning. As the world grapples with the environmental impact of industrial production, traditional techniques that harness local, eco-friendly materials offer valuable lessons. By sharing and adapting these practices, artisans can contribute to a global shift toward more sustainable forms of production.

The story of craftsmanship in the 21st century is, at its heart, a tale of convergence. It's a narrative that underscores the idea that while our tools, techniques, and materials may differ, the drive to create, to express, and to connect is a universal trait that transcends cultural boundaries.

As we look to the future, the role of cross-cultural exchanges in shaping the craftsmanship landscape is undeniable. In blending the old with the new, the local with the global, craftsmen are not only preserving their heritage but also laying the foundations for a dynamic and inclusive craft culture.

Thus, the journey of craftsmanship, fuelled by cross-cultural exchanges, is an ongoing adventure. It's a testament to the resilience and adaptability of artisans who, faced with an ever-changing world, continue to draw from a diverse palette of cultural influences to enrich their craft and, by extension, the world around them.

In conclusion, the cross-cultural exchanges that define modern craftsmanship are more than just a testament to global interconnectedness. They represent a mosaic of human creativity and ingenuity, a reminder that in our diversity lies our greatest strength. As we embrace these exchanges, we not only enhance our crafts but also weave tighter the bonds that connect us all, crafting a future where tradition and innovation walk hand in hand.

Fusion of Techniques and Styles

In today's world, the boundaries that once separated distinct styles and techniques in craftsmanship have become increasingly blurred. This fusion is not merely a reflection of individual creativity but a testament to the profound impact global cultures have had on the artisan world.

The exchange of ideas across continents, facilitated by the digital era, has ushered in an epoch of innovative approaches to traditional crafts. Artisans are no longer confined to the methods and materials endemic to their geographic location. Instead, they harness the power of global connectivity to blend techniques from disparate cultures, creating works that are as diverse in inspiration as they are in execution.

This amalgamation is perhaps most vividly seen in the realm of textile arts. Here, we find intricate Japanese Shibori dyeing techniques being merged with Scandinavian minimalistic designs. The result? Fabrics that speak a universal language of aesthetic appeal while retaining the essence of their origins.

Similarly, in woodworking, craftspeople might combine the intricate marquetry of the Middle East with the robust forms of Western furniture. This intersection not only challenges the craftsperson's skill set but also brings forth pieces that are a narrative of cross-cultural dialogue.

The influence of technology in this fusion cannot be overstated. Digital platforms serve as repositories of global inspiration; 3D printing and CNC machining allow for precision in bringing complex designs to fruition. Such synergy between handcraft and tech is generating a new breed of digital artisans who are redefining what it means to be a craftsperson in the 21st century.

However, this blend of styles and techniques does more than enrich the craft landscape. It acts as a bridge between cultures, fostering a deeper understanding and appreciation of diversity. Through their creations, artisans tell stories of cultural convergence, encouraging an ethos of global unity.

Moreover, this fusion is an avenue for sustainability in craftsmanship. By integrating materials and methods from various cultures, artisans are innovating more eco-friendly practices. This not only revitalises traditional crafts with a sustainable focus but also appeals to the modern consumer's growing environmental consciousness.

The educational domain has taken note of this trend, with institutions now offering courses that explore global craft techniques. This academic recognition not only legitimises the fusion as a field of study but also prepares future craftsmen for a globalised craft market.

Yet, as much as fusion enriches, it also challenges. The integrity of traditional crafts must be preserved even as they evolve. Artisans tread a fine line between innovation and the dilution of heritage. Respecting the origins and meanings of traditional techniques is paramount in this creative process.

This respectful blending of crafts from different cultures also sparks a renewed interest in heritage techniques that may be at risk of fading into obscurity. It grants them a new lease of life, embedding them in contemporary works that speak to a broad audience.

Innovation in fusion also extends to how craftsmanship is taught and passed down through generations. Crafting workshops and online platforms are becoming increasingly popular for sharing knowledge across borders, enabling an even wider exchange of techniques and styles.

Ultimately, the fusion of techniques and styles in craftsmanship is a vibrant celebration of human creativity and cultural diversity. It offers a compelling vision of a world where art transcends boundaries to unite us. It serves as a powerful reminder of our interconnectedness and the shared human spirit that finds expression through craft.

As we move forward, this fusion approach will undoubtedly continue to evolve, shaped by the tastes, technologies, and socio-political contexts of the times. Artisans and enthusiasts alike must remain open to these changes, embracing the endless possibilities that come with the cross-pollination of cultures.

In conclusion, the fusion of techniques and styles is not just a trend but a profound movement towards a more inclusive and innovative future in craftsmanship. It harnesses the best of what the world has to offer, channelling it into creations that are as meaningful as they are beautiful. In this endeavour, every craftsperson plays a pivotal role, whether as a custodian of tradition, an innovator, or both.

Therefore, let us all embrace this fusion, for in doing so, we not only enrich our own lives but contribute to a richer, more diverse world heritage of craftsmanship.

Chapter 14:
The Future of Handmade: Quality over Quantity

In an era where mass production and consumerism dominate, the allure of handmade, quality craftsmanship is experiencing a renaissance that beckons us towards a more sustainable and meaningful approach to consumption. As we delve into the essence of what it truly means to prioritise the integrity of craft over the allure of quantity, it's clear that a profound shift is underway. This transformative journey, illuminated by the growing demand for artisan products and the appeal of customisation, heralds a new age where the tactile connection to the objects we own and the stories they carry enriches our lives. Far from merely a revival of past traditions, this movement represents a bold leap into the future of craftsmanship—a future where the fusion of age-old skills with innovative practices ensures that the heritage of making, marked by the indelible touch of the craftsman's hand, not only endures but thrives. It is within this exciting paradigm that individuals across generations are finding fulfillment and purpose, driven by a shared belief in the value of one-of-a-kind creations that embody the essence of quality over quantity.

The Demand for Artisan Products

In today's fast-paced world, where mass production and standardisation have become the norm, a quiet but powerful shift is occurring. There's a growing appetite amongst consumers for products that tell a story, objects that carry the imprint of individual craftsmanship and the aura of the handmade. This chapter delves into the burgeoning demand for artisan products, understanding its roots and contemplating its implications for the future.

The resurgence of interest in artisanal goods isn't just a fleeting trend but a paradigm shift in consumer behaviour. Millennials and Gen Z, in particular, are at the forefront of championing this movement. They're not just looking for goods; they're seeking authenticity, sustainability, and a personal connection to the items they purchase. It's not merely about possession but about participation in a story that resonates with their values and aspirations.

Artisan products stand out in the market for their quality. The attention to detail, the use of superior materials, and the skillful techniques employed result in items that not only last longer but also perform better. This emphasis on quality over quantity is reflective of a broader societal shift towards conscious consumption and a departure from the disposable culture that has dominated for decades.

The allure of customisation also plays a significant role in the demand for artisanal goods. In a world where most products are designed for the masses, the opportunity to own something unique is highly prized. Artisans offer the chance to tailor products to individual tastes and needs, turning each piece into a personal statement.

Furthermore, the story behind each artisan product adds a layer of value that mass-produced items can't replicate. Knowing who made an item, where it came from, and the process behind its creation adds a level of connection and meaning. This narrative is a powerful draw for consumers who are increasingly interested in the origins and ethics behind what they buy.

This demand has significant implications for artisans and the future of craftsmanship. It presents an opportunity for craftsmen to connect directly with their audience, share their stories, and grow their businesses. The digital age has facilitated this connection, enabling artisans to reach a global market and find a niche audience passionate about their craft.

However, the rising demand also presents challenges. The market for artisan products is becoming increasingly saturated, making it harder for individual craftsmen to stand out. The pressure to produce at a pace that meets demand while maintaining the integrity of the handmade process can be overwhelming.

To navigate these challenges, artisans must leverage the very qualities that set them apart. Emphasising the uniqueness of their products, the sustainability of their practices, and the personal stories behind their work can help them differentiate themselves in a crowded market.

Education and skill development are also crucial. Aspiring artisans must be equipped not only with the technical skills of their craft but also with knowledge of business, marketing, and digital tools. This holistic skill set is essential for building a viable career in the modern landscape of craftsmanship.

Partnerships and collaborations offer another avenue for growth. By working together, artisans can expand their reach, pool resources, and create innovative products that push the boundaries of their craft. These collaborations can also foster a sense of community and support among craftsmen, countering the isolation that can sometimes accompany the artisanal journey.

The demand for artisan products has implications beyond the individual artisan or consumer. It's part of a larger movement towards sustainability and ethical consumption. By choosing handmade, consumers are supporting smaller businesses, reducing environmental impact, and promoting fair labour practices. This choice reflects a growing awareness and concern for the social and environmental consequences of consumption.

Despite the challenges, the future for artisan products looks bright. As consumers continue to seek out goods that reflect their values and aspirations, the demand for handcrafted, quality items is set to grow. This trend has the potential to revitalise traditional trades, foster innovation in craftsmanship, and contribute to a more sustainable and equitable economy.

In conclusion, the demand for artisan products is much more than a preference for the handmade. It's a reflection of changing values, a desire for connection, and a commitment to quality and sustainability. For artisans, responding to this demand requires not only craftsmanship but also innovation, adaptability, and a deep understanding of the stories and values that resonate with their audience.

As we move forward, the role of the artisan in society is being redefined. No longer seen as relics of the past, craftsmen are emerging as pioneers of a sustainable, ethical, and deeply personal approach to consumption. In embracing this role, artisans have the opportunity to not only sustain their craft but also shape the future of how we make, buy, and appreciate goods.

In considering the demand for artisan products, we're not just looking at a market trend but at a cultural shift that reflects deeper changes in how we view work, consumption, and community. It's a shift that challenges us to think differently about value, quality, and the power of the handmade, in crafting not just objects but a vision for the future that honours the best of tradition while embracing innovation and change.

The Lure of Customisation

In the ever-evolving landscape of the modern world, the allure of customisation in the realm of craftsmanship cannot be overstated. It signifies not just a shift in consumer preferences, but a deeper, more profound change in the way we interact with the objects that populate our lives. Customisation, at its core, is about personalisation and individuality, offering a stark contrast to the mass-produced goods that have dominated the market for decades.

The appeal of customisation is multifaceted. For one, it enables consumers to express their unique identity and style through the items they choose to possess. In a world where individualism is increasingly celebrated, having something made just for you is a powerful statement. It's about owning something that others don't have, or at least, not in the exact same way you do.

From the perspective of artisans and craftsmen, customisation opens up new avenues for creativity and innovation. It challenges them to not only master their craft but also to push the boundaries of what is possible. Each customised piece is a new adventure, a puzzle to solve, ensuring that their work remains dynamic and ever-evolving.

Moreover, customisation allows for a deeper connection between the craftsman and the consumer. It's a collaborative process, where the desires of the consumer are met with the expertise and artistic vision of the craftsman, culminating in a product that is imbued with meaning and personal significance. Such interactions not only enhance customer satisfaction but also instil a sense of pride and accomplishment in the craftsmen.

Technological advancements have played a pivotal role in facilitating this trend towards customisation. Digital tools and platforms have made it easier for craftsmen to collaborate with clients from anywhere in the world, breaking down geographical barriers and expanding their market. Technologies like 3D printing allow for intricate designs and patterns that would be difficult, if not impossible, to achieve manually.

This shift towards customisation also reflects a broader societal move towards sustainability and mindful consumption. In a world awash with excess, customised items offer an antidote to the disposable culture that has led to environmental degradation. They are often crafted with greater care, using higher quality materials, and with an eye towards longevity rather than obsolescence.

The lure of customisation goes beyond mere aesthetics or a preference for personalised items. It is emblematic of the changing values of consumers, particularly among Millennials and Gen Z, who favour authenticity, quality, and sustainability. This demographic is more willing to invest in products that reflect their values and ethos, and customisation perfectly aligns with this shift.

Education and skill development play a crucial role in supporting this move towards customisation. Aspiring craftsmen and artisans need to be well-versed not only in traditional techniques but also in the latest technologies that can enhance their craft. Trade schools and online learning platforms are increasingly incorporating these elements into their curriculum, preparing the next generation of craftsmen to meet the growing demand for customised goods.

Interestingly, the trend towards customisation is not limited to any specific domain but is evident across a broad spectrum of crafts, from furniture and jewellery to clothing and beyond. Each field is experiencing its own renaissance of sorts, as both consumers and craftsmen explore the possibilities that customisation brings.

The economic implications of this trend are significant. Customisation often comes at a premium, reflecting the additional time, skill, and effort required to produce personalised items.

However, consumers appear more than willing to pay this premium for goods that carry greater meaning and personal relevance. This willingness opens up lucrative opportunities for artisans and craftsmen who can cater to this demand.

The cultural impact of customisation should not be underestimated either. By fostering a culture that values quality over quantity and individual expression over uniformity, customisation is helping to redefine our relationship with the material world. It's a return to a more thoughtful and intentional mode of consumption, where each object tells a story and serves a purpose beyond its mere functional use.

However, the movement towards customisation is not without its challenges. Scaling customisation while maintaining quality and affordability is a conundrum many craftsmen face. Moreover, the need for continuous innovation and adaptation to client demands can be demanding, both creatively and physically. Yet, these challenges are also what make customisation such a thrilling and rewarding endeavour.

In conclusion, the lure of customisation in the world of craftsmanship is a testament to the changing priorities and values of society. It celebrates individuality, fosters sustainability, and champions authenticity, offering a compelling vision for the future of handmade goods. As we move forward, it is clear that customisation will play an integral role in shaping the trajectory of modern craftsmanship, making it an exciting time for both craftsmen and consumers alike.

One can't help but be inspired by the possibilities that lie ahead. The future of craftsmanship, powered by the dual engines of tradition and innovation, beckons us to explore, create, and ultimately, to redefine what it means to make and own things in the 21st century. The lure of customisation, with its promise of quality over quantity, is not just a trend; it's a movement towards a more personalised, sustainable, and meaningful consumer culture.

Indeed, as we journey into this future, it's crucial that both consumers and craftsmen continue to embrace change, seek out innovation, and celebrate the unique blend of tradition and technology that customisation embodies. In doing so, we not only enrich our own lives but also contribute to a more vibrant, sustainable, and creative world.

Chapter 15: Case Studies: Millennials & Gen Z Making an Impact

In a landscape where tradition and innovation converge, a new generation of artisans emerges. This chapter delves into real-life examples of Millennials and Gen Z individuals who've significantly impacted the craftsmanship domain, highlighting their journey in modernising traditional trades and pioneering tech-based crafts. These young visionaries, equipped with a passion for sustainability and technology, have seamlessly woven their ideals into the fabric of their crafts. They've not only embraced the essence of handiwork from generations past but have also catapulted it into the future through innovative practices. From turning eco-conscious ideas into profitable businesses to leveraging cutting-edge technology like 3D printing and augmented reality in their creations, their stories are a testament to the potential of youth in redefining the boundaries of artisanship. These case studies serve as a beacon, inspiring readers to view the fields of craftsmanship not as relics of a bygone era but as vibrant platforms for change and creativity. The journey of these young craftsmen underlines a crucial message: with the right blend of vision, innovation, and reverence for tradition, it's possible to forge a career that's not only fulfilling but also future-proof, showcasing craftsmanship as a dynamic and evolving field.

Innovators in Traditional Trades

In a world that's rapidly advancing technologically, it's easy to assume that traditional trades might fall by the wayside, rendered obsolete in the face of digitalisation. However, a fresh wave of innovators, primarily from the millennial and Gen Z cohorts, are proving this assumption wrong. They're not only embracing traditional crafts but are also integrating them with modern technology to create something truly unique and forward-thinking. This chapter delves into how these young craftsmen are making a significant impact, transforming and sustaining traditional trades for the future.

At the heart of this revolution are individuals who respect the heritage of craftsmanship but are not afraid to experiment with new methods. One such example is the integration of 3D printing into pottery and ceramics. This fusion enables artisans to create designs that would be impossible or incredibly time-consuming by hand, opening up new realms of creativity and functionality. The marriage of 3D printing with clay has ushered in a new era for ceramicists, combining the tactile joy of traditional pottery with the precision and versatility of modern technology.

Another area where innovation is thriving is in textiles. Millennials and Gen Z are using digital fabrication methods, such as laser cutting and digital embroidery, to bring unprecedented intricacy and personalisation to their creations. These technologies allow for the replication of incredibly detailed designs that pay homage to traditional patterns while pushing the boundaries of what's possible in fabric manipulation.

Furniture making, too, has seen a renaissance under the touch of young innovators. By incorporating sustainable materials and using computer-assisted design (CAD) software, these new-age craftsmen can produce durable, bespoke pieces with a fraction of the environmental impact. The blend of sustainability with traditional woodworking techniques stands as a testament to how the past and future can coexist harmoniously in the realm of craftsmanship.

Moreover, in the sphere of metalworking, the utilisation of programmable machines and robotics in conjunction with age-old forging techniques is creating pieces of both functional and artistic merit. This approach not only increases efficiency and precision but also opens the door to new forms and textures that were once beyond reach.

Preservation of traditional crafts is also a significant focus for these young artisans. Around the globe, millennials and Gen Z are leveraging social media and online platforms to share their skills and knowledge, ensuring that these ancient practices are not lost to time. They're teaching and inspiring others, demonstrating that traditional trades have a place in the modern world and can offer rewarding, sustainable careers.

This resurgence and transformation of trades aren't just about preserving the past; it's about creating a vibrant, dynamic future. Young innovators are at the forefront, demonstrating that with creativity and technology, traditional crafts can evolve to meet contemporary needs and sensibilities without losing their soul.

One illuminating example is the craft of glassblowing, an ancient art that has remained largely unchanged for centuries. Innovators in this field are experimenting with eco-friendly materials and energy-efficient furnaces, making the process more sustainable while still producing breathtakingly beautiful pieces.

In the culinary arts, there's a movement towards combining traditional fermentation techniques with modern culinary science to create innovative and health-conscious food items. This blend of old and new approaches is leading to the development of flavours and textures that are completely novel, yet rooted in tradition.

The realm of jewellery making has also seen transformative changes, with artisans employing CAD and 3D printing to design intricate pieces that reflect both personal and cultural narratives. These technologies allow for customisation on a level that was previously impossible, offering consumers the chance to own pieces that are truly unique.

Bookbinding is another craft that's being revolutionised by young innovators. By integrating traditional binding techniques with modern designs and materials, these craftsmen are ensuring that the art of bookbinding remains relevant in the digital age, appealing to collectors and bibliophiles who cherish the tactile experience of reading.

The dedication of these young innovators to their crafts goes beyond mere production; it's about building communities, sharing knowledge, and ensuring sustainability. They're not just making things; they're making a difference, proving that traditional trades can adapt, survive, and thrive in the modern world.

In essence, the integration of technology and innovation into traditional trades by millennials and Gen Z is breathing new life into ancient crafts, ensuring their relevance and sustainability. This chapter has showcased just a fraction of the incredible work being done across various fields, highlighting the limitless potential when tradition meets innovation.

The journey of these young craftsmen is a beacon for future generations, demonstrating that with passion, creativity, and respect for both heritage and progress, it's possible to carve out fulfilling careers that contribute positively to our culture and economy.

As we look towards the future, it's clear that the landscape of craftsmanship is evolving in exciting and unexpected ways. The innovators in traditional trades are leading this charge, ensuring that the rich tapestry of human creativity and skill continues to flourish, bridging the gap between our past and our future. Their work is not just a celebration of craftsmanship; it's a blueprint for a world where tradition and innovation coexist in harmony, enriching our lives in myriad ways.

Pioneers in Tech-Based Crafts

In the vibrant tapestry of modern craftsmanship, a fascinating narrative is emerging, one that seamlessly stitches the rich heritage of traditional trades with the boundless possibilities offered by technology. This narrative is being written by a new generation - Millennials and Gen Z - who are not just passively inheriting traditions but are actively reimagining them for the digital age. They serve as the pioneers of tech-based crafts, a testament to the potential that lies at the intersection of age-old skills and cutting-edge technology.

Let's take, for instance, the realm of textile design. Gone are the days when production was solely dependent on manual looms. Today, we witness young artisans melding the tactile joy of textile creation with the precision and scale offered by digital fabrication techniques. They employ computer-aided design (CAD) software to bring intricate patterns to life, harnessing the power of digital looms to produce textiles that speak volumes of our contemporary aesthetic while paying homage to traditional motifs.

Similarly, the domain of ceramics has witnessed a radical transformation. The wheel and kiln, while still central to the craft, are now being complemented with 3D printing technologies. This confluence of old and new enables artisans to experiment with forms and textures that were previously unimaginable, pushing the boundaries of what can be achieved with clay.

In the sphere of metalwork, we observe a similar trend. Young craftsmen are employing laser cutting and CNC machining to craft intricate designs with a level of precision that hand tools could never achieve. This technological leverage affords them the opportunity to experiment with new materials and techniques, heralding a renaissance in metal crafts that fuses traditional skills with futuristic visions.

Woodworking, too, has been revolutionised. The integration of digital fabrication tools has opened up new possibilities for customisation and creativity. CNC routers, for example, allow for the creation of complex, precise cuts that elevate the craftsmanship and functionality of wooden products. This merging of digital precision with the tactile beauty of wood results in pieces that are not only aesthetically pleasing but also embody the spirit of innovation that defines our era.

But the influence of technology on crafts doesn't end with production; it extends to the realm of knowledge sharing and community building as well. Online platforms have become vibrant hubs where young artisans from across the globe exchange ideas, techniques, and inspirations. This digital fellowship is not only accelerating the spread of traditional skills but is also fostering a culture of collaboration and innovation that transcends geographical boundaries.

Take the story of a young jeweller who uses computer-aided design (CAD) to craft bespoke pieces that incorporate both futuristic designs and elements of ancient jewellery-making traditions. This blend of the old and the new, the traditional and the technological, resonates deeply with consumers seeking pieces with a story, a history, and a future.

Or consider the artisans who are breathing new life into the craft of glassblowing by integrating LED lights and electronic sensors into their creations. These luminous sculptures not only challenge our perceptions of what glass art can be but also serve as a symbol of the potential for technology to enhance, rather than replace, the human touch in crafts.

In the context of sustainability, technology has been a boon for young craftspeople committed to eco-friendly practices. They are crafting products from recycled materials using technologically advanced methods that minimise waste and energy consumption. This commitment to sustainability is not just about preserving the environment but also about crafting a future where our material culture is in harmony with the planet.

The fusion of technology and traditional crafts is also creating new pathways for entrepreneurship. Armed with social media tools and e-commerce platforms, young artisans are reaching audiences far beyond their local markets. They're not merely selling products; they're telling stories, sharing processes, and building brands that reflect a new ethos of craftsmanship that is inclusive, innovative, and informed by a global perspective.

This technological renaissance in crafts is not without its challenges. The learning curve for mastering both traditional skills and new technologies can be steep. Access to expensive equipment and software may also pose barriers for some. Yet, despite these hurdles, the drive among Millennials and Gen Z to innovate within their trades is palpable. They are leveraging scholarships, online tutorials, and community workshops to bridge the gap, demonstrating a resilience and resourcefulness that is truly inspiring.

One lesson that stands out from these pioneers is the importance of adaptability. They show us that to thrive in the evolving landscape of craftsmanship, one must be willing to learn continuously, experiment bravely, and embrace change eagerly. They are moving away from the notion of the solitary artisan, working in isolation, to a more collaborative and interconnected approach, making craftsmanship a dynamic and collective endeavour.

As we reflect on the contributions of these young artisans, it's clear that the future of craftsmanship is one where tradition and innovation coexist in harmony. It's a future that promises not only aesthetic and material richness but also a deep sense of purpose and sustainability. Indeed, the pioneers of tech-based crafts are not just making objects; they're shaping the very fabric of our culture, ensuring that the age-old spirit of craftsmanship not only survives but thrives in the digital age.

This chapter serves as an ode to their ingenuity and passion. As we turn the pages of their stories, we're not just learning about techniques and technologies; we're witnessing the dawn of a new era in craftsmanship. An era that champions diversity, embraces technology, and values sustainability. An era that promises to inspire future generations of craftsmen to continue the journey of innovation, with their roots deeply entrenched in tradition but their sights set firmly on the horizon.

In celebrating these pioneers in tech-based crafts, we're reminded of the infinite possibilities that unfold when human creativity is coupled with technological advancement. It's a powerful testament to the resilience, adaptability, and visionary spirit of Millennials and Gen Z, who are redefining what it means to be an artisan in the 21st century. As they chart new territories, blending the old with the new, they light the way for a future where craftsmanship continues to enrich our lives in ways we can only begin to imagine.

Chapter 16: Challenges Facing Young Craftsmen Today

In an era where innovation intertwines with tradition, young craftsmen find themselves navigating a complex landscape filled with both unprecedented opportunities and formidable challenges. Among the myriad of obstacles, financial hurdles prominently stand, with the cost of materials, tools, and education often proving to be steep barriers to entry. Moreover, the quest for recognition and respect in a market saturated with mass-produced goods requires not just talent but also astute branding and marketing skills. Young artisans must carve their niche in an environment that frequently undervalues the time-honoured skills and meticulous labour involved in crafting bespoke items. Despite these trials, the spirit of craftsmanship persists, buoyed by a passionate community that values authenticity, sustainability, and the personal touch of handmade goods. This chapter delves into how emerging craftsmen are confronting these challenges head-on, employing innovation, leveraging technology, and fostering community to ensure the vibrancy and relevance of craftsmanship for generations to come.

Financial Hurdles

In the chapter that unfolds, we delve into one of the most formidable challenges young craftsmen face today – financial hurdles. These are not mere bumps on the road but significant barriers that can deter even the most passionate aspirants from pursuing a career in craftsmanship. Understanding these financial challenges is the first step toward overcoming them and paving a way for a new generation of crafters who can merge tradition with innovation in their trade.

Firstly, the initial investment required to embark on a craft career can be substantial. Quality tools, materials, and workspace rentals do not come cheap. For a young individual at the outset of their career, these expenses can seem insurmountable. Unlike digital domains where one primarily needs a computer and software, traditional crafts demand a variety of physical resources to begin with.

Furthermore, the cost of education and skill development further complicate matters. While online learning has made knowledge more accessible, mastering a craft often requires hands-on experience and guidance from seasoned artisans. This means enrolling in specialized trade schools or undertaking apprenticeships, which are time-intensive and sometimes costly due to tuition fees and living expenses.

Adding to the financial strain are the challenges associated with finding a market for artisan products. The process of establishing a brand, marketing one's work, and reaching potential customers requires not just creative skills but also an understanding of business and digital marketing strategies. For craftsmen just starting, these demands can be overwhelming and resource-intensive.

The competitive landscape in the artisanal sector also poses a significant financial hurdle. As the market grows, distinguishing oneself and one's work becomes increasingly challenging. This requires investment in unique materials, continuous skill development, and possibly even technological integration, all of which demand financial resources.

Moreover, the unpredictability of income in the early stages of a crafting career discourages many. Unlike conventional jobs offering a stable salary, earnings from craftsmanship can fluctuate greatly based on market demand, seasonality, and other factors. This makes financial planning and security a constant concern for emerging craftsmen.

Access to capital is another substantial barrier. Traditional banks and financial institutions often regard crafts and artisan businesses as high-risk ventures, making loans and financial support difficult to obtain. This lack of support stifles innovation and prevents talented individuals from pursuing their crafting ambitions.

There's also the challenge of scalability and the associated costs. Transitioning from a solo operation to a larger business setup involves significant financial outlay in production, labour, and distribution channels. Without substantial investment or backers, scaling an artisan business can seem like a distant dream for many young craftspeople.

On a more positive note, overcoming these financial hurdles is not impossible. A combination of creative problem-solving, leveraging digital platforms for education and marketing, and engaging in collaborative and community-based initiatives can open up new avenues for financial sustainability in craftsmanship.

Grants, scholarships, and fellowships geared towards artisans can alleviate some of the educational and developmental costs. Seeking out these opportunities requires diligence and persistence but can provide much-needed support for ambitious craftsmen.

Digital marketplaces and social media platforms offer low-cost, high-reach options for marketing and selling artisanal products. By building a strong online presence, craftsmen can connect with a global audience without incurring the hefty costs traditionally associated with marketing and distribution.

Crowdfunding platforms present another viable avenue for raising capital. By presenting their vision and products to potential backers, craftsmen can secure the financial backing they need to kickstart or scale their operations, bypassing traditional funding barriers.

Collaborations and partnerships with other artisans or businesses can also help in mitigating financial risks. Sharing workspace, materials, and even sales platforms can reduce individual costs while fostering a spirit of community and mutual support among craftsmen.

Embracing technology and innovation not only in craft creation but also in business practices can significantly reduce operational costs and open up new revenue streams. From employing 3D printing to reduce material waste to selling digital templates or tutorials, there are myriad ways in which modern technology can complement traditional craftsmanship.

In conclusion, while financial hurdles are undeniably substantial, they are not insurmountable. With a blend of traditional skills, modern technology, and innovative approaches to business, young craftsmen can overcome these challenges. The path may be fraught with obstacles, but it is also ripe with opportunity for those willing to navigate it with creativity and determination. Encouraging a new generation of craftsmen to tackle these hurdles head-on, with resilience and innovation, not only secures their future but also preserves the rich heritage of craftsmanship for future generations.

The Struggle for Recognition and Respect

In an era where every profession seems to be vying for attention and validation, young craftsmen are facing a particularly steep uphill battle for recognition and respect in their chosen fields. This struggle is multifaceted, influenced by societal trends, economic factors, and the evolving landscape of technology.

Traditionally, craftsmanship was revered as a noble pursuit, with artisans holding a respected place in society. Their skills, honed over years, if not decades, were indispensable to the daily lives and cultural expressions of their communities. Today, however, as mass production and digital technologies become increasingly predominant, the intimate connection between a craftsman and their work can seem out of step with the current zeitgeist.

For many young artisans, this disconnect is not merely academic; it manifests in tangible barriers to their professional development. At the heart of the matter is the pervasive undervaluation of skilled manual labour in favour of white-collar professions perceived as more prestigious or financially rewarding. This societal bias can discourage young talents from pursuing craftsmanship as a viable career path.

Moreover, the apprenticeship model, once the backbone of craftsman education, faces its own modern-day challenges. In a world where speed and efficiency are often prioritised over depth and quality, the slow, meticulous process of mastering a craft can seem outdated. Consequently, finding mentors willing to invest time in the next generation is becoming increasingly difficult.

Compounding these issues is the financial reality for many aspiring artisans. The initial stages of a craftsmanship career often require significant investment in tools, materials, and education, with no immediate guarantee of return. This economic barrier can be insurmountable for some, particularly when juxtaposed with the allure of more immediately lucrative career paths.

Yet, despite these challenges, the pursuit of craftsmanship remains a deeply compelling path for many. This is where the role of recognition and respect becomes crucial. Acknowledgement of their skills and contributions serves not just as a moral support but also opens doors to opportunities, collaborations, and financial stability.

In response to the struggle for visibility, many young craftsmen are turning to social media and online platforms to showcase their work. The digital world offers a global stage, but breaking through the noise and capturing the attention of a wider audience requires savvy, persistence, and a touch of creativity.

Furthermore, the evolving landscape of craftsmanship means integrating new technologies and sustainable practices into traditional methods. Young artisans who navigate this fusion successfully often find that it elevates their work, garnering respect from both traditionalists and innovators. Yet, achieving this synthesis is no small feat and underscores the need for a supportive and appreciative community.

From an educational standpoint, there's a growing recognition of the importance of revamping curriculums to reflect the changing realities of craftsmanship. By providing training that encompasses both traditional techniques and modern innovations, institutions can better prepare students for the challenges ahead. This also serves to elevate the perceived value of craftsmanship education in the broader societal context.

At the industry level, there's a rallying call for established craftsmen and stakeholders to advocate for recognition and respect for the craft. This involves pushing for fair compensation, opportunities for showcasing work, and the creation of platforms for knowledge exchange and mentorship. Only through collective effort can the barriers facing young artisans be dismantled.

Another key aspect of the struggle for recognition is the need for public education and awareness. Dispelling myths about craftsmanship, highlighting the sector's economic and cultural significance, and celebrating success stories can all contribute to a shift in perception. The more the public understands and values the work of craftsmen, the greater respect and recognition they will command.

In this respect, festivals, fairs, and exhibitions play a vital role. By providing artisans with a physical platform to display their work, connect with peers, and engage directly with consumers, these events can foster a greater appreciation for the skill, creativity, and dedication inherent in craftsmanship. They also serve as important networking opportunities, opening doors to collaborations and commissions.

For young craftsmen, persistence in the face of these challenges is crucial. Amidst the struggle for recognition and respect, it's important to remember that the value of their work transcends current trends and market dynamics. By staying true to their passion and continuously striving for excellence, they not only enhance their skills but also contribute to the rich tapestry of cultural heritage.

Ultimately, the struggle for recognition and respect in the realm of modern craftsmanship is a reflection of broader societal shifts. As we move towards an increasingly digital and automated future, the importance of preserving manual skills and artistic expression becomes ever more poignant. In championing young craftsmen today, we not only ensure the survival of age-old trades but also enrich our collective culture and identity for generations to come.

It's a challenging path that young artisans have chosen, but it is also uniquely rewarding. In their hands lies the potential to redefine craftsmanship for the 21st century, merging tradition with innovation to create something truly exceptional. Their success hinges not only on their individual skills and dedication but also on the collective effort to recognise and respect the invaluable contribution of craftsmen to our world.

Chapter 17: The Politics of Craftsmanship: Regulation and Advocacy

In the labyrinth of tradition and innovation, the politics of craftsmanship emerges as a pivotal challenge for today's artisans, echoing the struggles and victories of navigating bureaucracy and advocating for meaningful change. As craftsmen blend the old with the new, they find themselves at odds with regulations that often lag behind the pace of technological advancement and societal needs. This chapter delves into the intricate dance between maintaining standards and pushing the boundaries of what is possible, highlighting the vital role of advocacy in crafting the future. Through engaging narratives, we explore how craftsmen are lobbying for changes that not only permit but encourage innovation, ensuring that ancient skills flourish in a modern context. It's a testament to the power of collective action and vision in moulding a regulatory environment that fosters creativity, sustainability, and inclusivity. As we delve into the strategies that have led to breakthroughs in policy and perception, it becomes clear that the spirit of craftsmanship is not only about the mastery of skills but also about influencing the frameworks within which they operate. This chapter serves both as a call to action and a beacon of hope for those navigating the politics of craftsmanship, offering insights into how persistence, collaboration, and a deep understanding of one's craft can lead to far-reaching impacts beyond the workshop.

Navigating Bureaucracy

In the quest to harmonise innovation with tradition, craftsmen of all calibres face an array of bureaucratic challenges. These hurdles, often seen as intimidating barriers, can significantly impact the journey of a craftsman from concept to market. However, understanding and navigating these bureaucratic processes effectively can turn potential obstacles into stepping stones towards success.

The journey begins with acknowledging the spectrum of regulations that impact craftsmanship. From health and safety standards to trade permits and intellectual property laws, the list is exhaustive. Each of these regulations serve a purpose, aiming to ensure quality, protect consumers, and foster fair competition. Yet, the complexity and often the sheer volume of such regulations can seem daunting, especially to newcomers in the world of craftsmanship.

One common thread that weaves through the tapestry of bureaucracy is the requirement for permits and licenses. Whether it's a small at-home studio producing ceramic pots or a tech-savvy artisan specialising in 3D-printed jewellery, understanding what permits are necessary and how to obtain them is crucial. Often, these permits are designed to ensure that the business operations comply with local laws and standards, which can vary significantly from one jurisdiction to another.

The role of trademarks and copyrights cannot be understated in the realm of craftsmanship. As artisans pour their heart and soul into creating unique pieces, protecting these creations from imitation or outright theft is paramount. However, the process of registering a trademark or copyright can be laden with jargon and legal complexities. Seeking professional advice or utilising resources specifically designed for artisans can demystify this process, providing much-needed protection for one's creations.

Similarly, navigating the labyrinth of health and safety regulations is another critical aspect of bureaucracy that can't be overlooked. These regulations ensure that workshops and products are safe for both the creators and the consumers. Compliance not only minimises the risk of accidents and health issues but also enhances the reputation of the craftsman as a responsible maker.

Another dimension of bureaucracy that often goes unnoticed until it becomes a pressing issue is zoning laws. These laws dictate where certain types of business activities can take place. For craftsmen, this could mean restrictions on operating a workshop in a residential area or specific requirements for retail spaces. Early consultation with zoning authorities can prevent costly missteps and ensure that the physical space aligns with legal requirements.

Environmental regulations also play an increasingly significant role in modern craftsmanship. With a growing emphasis on sustainability and eco-friendly practices, complying with environmental laws is not just about adherence but also about embracing a philosophy that resonates with consumers. From waste management to the use of hazardous materials, understanding these regulations is essential for building a brand that's both ethical and compliant.

Amid these challenges, the concept of advocacy and collective action emerges as a beacon of hope. By joining trade associations or craft guilds, artisans can find support and resources to navigate the bureaucratic maze. These organisations often offer guidance, workshops, and even legal assistance to their members, making the journey less solitary and more manageable.

Advocacy goes beyond individual benefit, extending to lobbying for changes in regulations that reflect the evolving nature of craftsmanship. Through collective voices, craftsmen can influence policymakers to consider the unique needs and contributions of the crafts sector, leading to more supportive and relevant regulations.

Technology, too, plays a crucial role in demystifying bureaucracy. Online platforms and software offer tools for managing permits, copyright registrations, and compliance with regulations, simplifying what once was a cumbersome process. These technological solutions not only save time but also provide artisans with more opportunities to focus on their craft.

Financial regulation is another area where craftsmen need to tread carefully. Understanding tax obligations, import-export regulations, and financial reporting requirements is crucial for running a sustainable business. Here, professional advice from accountants familiar with the crafts industry can be invaluable, ensuring that artisans stay compliant while optimising their financial performance.

Insurance is another key consideration, providing a safety net against unforeseen circumstances. From property damage to liability claims, having the right insurance coverage can safeguard both the business and its customers. Navigating insurance options and understanding what coverage is necessary can protect artisans from significant financial risks.

Education and awareness are, ultimately, the best tools for navigating bureaucracy. Participating in seminars, leveraging online resources, and engaging with a community of fellow craftsmen can illuminate the path through regulatory requirements. Knowledge empowers artisans to make informed decisions, turning bureaucratic challenges into manageable aspects of their business.

In conclusion, while bureaucracy can seem like a daunting aspect of modern craftsmanship, it is an integral part of ensuring quality, safety, and fairness in the industry. By understanding and effectively navigating these regulations, artisans can not only protect their livelihood but also elevate their practice. Through advocacy, technology, and education, the bureaucratic labyrinth becomes less intimidating, opening up a world of possibilities for craftsmen eager to merge tradition with innovation.

As we move forward, embracing change and innovation while respecting tradition, the ability to navigate bureaucracy skillfully will remain a critical skill for craftsmen. It's not simply about compliance; it's about building a foundation for success in an evolving landscape where craftsmanship continues to thrive, enriched by technology and driven by passion.

Lobbying for Change

In the intricate web of the modern world, where tradition intertwines with technology, the craftsperson finds themselves at a unique crossroads. It's a place where skill and passion meet the rigorous demand for innovation and sustainability. Yet, despite the vibrant fusion of old and new, craftsmen often grapple with regulatory landscapes that can stifle growth and creativity. It's within this context that the act of lobbying for change becomes not just pertinent but essential.

At its core, lobbying for change within the realm of craftsmanship is about advocating for policies that recognise and nurture the unique contribution of artisans to society and the economy. It's about creating a space where traditional skills are honoured alongside technological advancements, where regulations support rather than hinder the evolution of trades.

The journey begins with understanding the current legislative environment and its impact on craftsmanship. Many laws and regulations, established in times vastly different from our own, fail to account for the innovations in materials, processes, and marketing that modern artisans employ. This mismatch can lead to significant barriers for those looking to make a living through their trade.

To challenge and eventually remodel these outdated regulations, the crafting community must engage in meaningful dialogue with policymakers. This involves presenting clear, evidence-based arguments that highlight the economic, cultural, and educational value of craftsmanship in the 21st century. Artisans, educators, and industry professionals must come together, forming a united front to advocate for change.

One promising avenue is the formation of coalitions and trade associations that can represent craftsmen's interests at local, national, and even international levels. These organisations can serve as a powerful voice, one that's capable of influencing policy and fostering a more supportive environment for artisans.

Focusing on education and skill development is another critical aspect of lobbying for change. By highlighting the role of craftsmanship in fostering innovation, sustainability, and job creation, advocates can make a compelling case for the inclusion of trade skills in educational curriculums. This not only addresses the skills gap but also elevates the status of artisans within society.

Moreover, the narrative must shift to position craftsmanship as a cornerstone of the circular economy. Lobbying efforts should showcase how traditional trades, with their emphasis on quality, durability, and repairability, can lead to more sustainable consumption patterns. Artisans have a unique opportunity to lead by example, illustrating how their practices align with global sustainability goals.

In the digital age, protecting intellectual property becomes increasingly complex yet crucial for artisans. Engaging in lobbying to refine copyright laws to reflect the new realities of digital sharing and commerce is imperative. Ensuring that craftspeople's creations are safeguarded under the law not only protects their livelihoods but also encourages innovation and sharing of knowledge.

Technological innovations such as 3D printing, IoT, and VR/AR are redefining what it means to be a craftsman. Lobbyists must argue for policies that provide access to these technologies, alongside traditional tools and methods, ensuring that artisans can compete in the modern marketplace.

One cannot overlook the importance of accessibility and inclusion in the world of crafts. Lobbying for change also means fighting for policies that make trades accessible to all, regardless of gender, race, or socioeconomic status. This involves advocating for funding, support networks, and programs that break down barriers and foster a diverse community of craftspeople.

Success in these endeavours requires not just passion but patience. Changing policies is a marathon, not a sprint. It involves countless meetings, discussions, and sometimes setbacks. Yet, the potential rewards - a world where craftsmanship is fully valued and supported - are worth the effort.

The role of social media in this context cannot be understated. By harnessing the power of online platforms, the crafting community can raise awareness, garner support, and even pressure policymakers into action. It's a tool that, when used effectively, can amplify the voice of artisans like never before.

Integral to all these efforts is the need for craftsmen to stay informed and engaged with the political process. Voting, participating in consultations, and simply staying abreast of policy developments are all crucial steps in ensuring that the crafting community's needs are not overlooked.

Lastly, it's vital to celebrate the victories, no matter how small. Every regulation modified, every policy improved, and every artisan's voice heard represents progress. It's a testament to the power of advocacy and a reminder that change is not only possible but achievable.

In conclusion, lobbying for change in the landscape of craftsmanship is an ongoing journey, one marked by challenges but also immense possibility. Through united efforts, clear communication, and unwavering dedication, the crafting community can shape a future where tradition and innovation coexist harmoniously, and where the artisan's role is celebrated and preserved for generations to come.

Chapter 18: Collaborative Craftsmanship: Projects and Partnerships

In the stimulating world of today, collaborations across various disciplines and borders have become a beating heart for the evolution of traditional trades, redefined through the lens of modern technology and innovation. This chapter delves into how interdisciplinary projects and global partnerships are forging new pathways for craftsmen, enabling them to blend ancient wisdom with contemporary techniques, thus broadening their horizons beyond local confines. It's here that the spirited exchange of ideas and practices unveils its true potential, demonstrating that when diverse minds unite, they cultivate an extraordinary tapestry of creativity and excellence. By highlighting successful synergies between craftspeople from disparate backgrounds, we explore the boundless opportunities that such collaborations present, not only in enhancing the appeal and sustainability of trades but also in paving a robust foundation for future generations. This collective journey towards innovative craftsmanship doesn't just promise a revival of interest in trades long considered obsolete; it champions a future where craftsmanship is universally acknowledged as a vibrant and essential contributor to our cultural and economic well-being. As we navigate through stories of shared endeavours, it becomes evident that in unity, there exists a remarkable capacity to rejuvenate old worlds with new breaths of life, turning collective dreams into tangible realities.

Interdisciplinary Projects

In a world where the boundaries between disciplines become increasingly blurred, crafting a career that intertwines various fields of interest isn't just possible; it's becoming the norm. Interdisciplinary projects, standing at the confluence of art, science, technology, and craftsmanship, offer a fertile ground for innovation, demanding a new kind of craftsman: one who is not defined by a single skill but by the ability to learn, adapt, and meld diverse crafts into something entirely new and groundbreaking.

The beauty of interdisciplinary projects lies in their inherent nature to challenge and expand the traditional notions of craftsmanship. These projects are not just about bringing together different skills for the sake of diversity. They are about creating a synergy that transcends the sum of its parts, leading to outcomes that could not have been achieved through a single discipline. This approach is reshaping how we perceive and value the role of craftsmanship in contemporary society.

Consider the example of a project that combines woodworking with digital fabrication technologies like CNC routing and 3D printing. Here, the age-old art of shaping wood meets the precision and versatility of modern technology. The result? Exquisite pieces that maintain the warmth and uniqueness of handcrafted woodwork, yet exhibit complexities or functionalities that might be impossible to achieve through traditional methods alone.

Such collaborations do not merely rest on the laurels of combining physical skills with technology. They extend into the realms of sustainability and eco-friendliness, where traditional methods of crafting are re-evaluated to meet contemporary demands for sustainable practices. Projects integrating recycled materials with traditional crafting techniques reflect a commitment to eco-conscious craftsmanship, providing new life to discarded items while preserving artisanal heritage.

The integration of technology into traditional crafts also opens up exciting possibilities for bespoke designs and customisations at a scale previously unimaginable. Tailored solutions that cater to individual needs and preferences become feasible, combining the personal touch of traditional crafts with the efficiency and precision of modern technology. This customization extends beyond the product itself into the realm of user experience, where craftspeople can engage clients in the design process, making them co-creators of the final piece.

Educational projects bring another dimension to interdisciplinary collaboration, merging pedagogy with craftsmanship to develop innovative learning experiences. Through hands-on projects that blend crafting skills with subjects like mathematics, history, or science, students gain a deeper understanding of both the practical and theoretical aspects of their studies, fostering a holistic approach to education that prepares them for the challenges of the future.

In the realm of art, interdisciplinary projects challenge the boundaries between the artisan and the artist, blurring the lines that once separated functional crafts from 'high' art. This fusion not only elevates the perceived value of crafts but also encourages a cultural dialogue between tradition and modernity, opening up a space where crafts are appreciated not just for their utility but as forms of artistic expression.

Community-led projects further illustrate the power of interdisciplinary collaboration, uniting craftspeople, designers, technologists, and the wider community in initiatives aimed at revitalising local traditions, economies, or environments. Such projects underscore the role of craftsmanship in building and strengthening communities, proving that the act of making can be a powerful catalyst for social change and cohesion.

The digital realm, with its vast networks and platforms, plays a pivotal role in facilitating interdisciplinary projects. Online collaborations can transcend geographical limitations, allowing craftspeople to work on projects with partners located across the globe. This global exchange of ideas and techniques enriches the crafting process, fostering a new era of craftsmanship that is both locally grounded and globally connected.

One cannot discuss interdisciplinary projects without considering the challenges they pose, particularly in terms of communication and integration of different knowledge domains. The successful execution of such projects requires a level of openness, resilience, and willingness to step out of one's comfort zone, embracing the unknown and the unpredictable. Yet, it is precisely this complexity that makes interdisciplinary projects so rewarding, as they push boundaries and catalyse innovation.

The future of craftsmanship, thus, lies in its ability to morph and adapt, to forge new paths that bridge disparate disciplines. Aspiring craftsmen must therefore be voracious learners, adept in not just their own trade but also in how it can intersect with others. The interdisciplinary projects of today are crafting the blueprint for the future, one where the craftsperson is not just a maker, but a visionary, an innovator, and a creator without borders.

Supporting interdisciplinary projects also means re-evaluating educational and professional pathways, ensuring they provide the flexibility and interdisciplinary exposure needed to thrive in today's complex, interconnected world. Education systems, industry standards, and public policies must evolve to recognize and nurture the multifaceted skill sets required for such work.

In conclusion, interdisciplinary projects are much more than an amalgamation of different crafts; they are a testament to the limitless potential of human creativity and ingenuity. They encourage us to look beyond traditional boundaries, to reimagine what we are capable of creating together. In doing so, they pave the way for a future where craftsmanship continues to play a vital, vibrant role in society, enriching our lives in ways we have yet to imagine.

Global Collaborations

In a world that's more connected than ever, the essence of collaborative craftsmanship has transcended beyond local communities, embracing a global stage. This shift towards international collaboration is not just a trend but a powerful movement that has redefined the boundaries of traditional and modern craftsmanship. It's a testament to how craftsmen from diverse cultures and backgrounds are uniting their skills to create something unprecedented.

The advent of digital technology has been a significant catalyst in this evolution. Craftspeople, who once may have felt isolated in their studios or workshops, are now part of a global fabric. They're leveraging platforms and tools that enable them to share, innovate, and co-create with peers thousands of miles away. This digital rendezvous has not only broadened their horizons but has also enriched the tapestry of global craftsmanship with novel ideas and techniques.

Consider, for instance, a woodworker in Japan, known for his mastery in traditional joinery, collaborating with a contemporary furniture designer in Sweden. Together, they fuse minimalistic Scandinavian design with the intricate Japanese joinery, creating pieces that are both aesthetically appealing and structurally sound. Such partnerships are a celebration of cultural diversity and the shared human spirit of creativity.

Moreover, these global collaborations are a beacon of innovation in sustainability. Craftsmen worldwide are joining forces to tackle environmental challenges, brainstorming and implementing eco-friendly practices in their creations. By exchanging knowledge on sustainable materials and techniques, they're not just crafting for today but are also paving the way for a more sustainable future.

Education has also benefited greatly from this interconnectedness. Workshops and masterclasses that were once geographically bound can now reach a global audience through online platforms. This has democratized learning, allowing aspiring craftsmen in remote areas to access knowledge from master artisans around the world. It's a powerful reminder that in the realm of craftsmanship, barriers to learning are being dismantled, one collaboration at a time.

The economic landscape of craftsmanship has been revitalized through these global partnerships. Artisans are tapping into markets beyond their localities, finding new audiences for their crafts. This has not only helped in preserving traditional crafts that were once under threat of disappearing but has also provided financial stability to artisans, fostering a renewed sense of pride in their work.

Yet, amid this celebration of global collaboration, challenges persist. Language barriers, cultural differences, and logistical issues are just a few hurdles that craftsmen face. However, the spirit of collaboration often turns these challenges into opportunities for learning and growth. By navigating through these complexities, craftsmen emerge more adaptable, resilient, and open-minded.

The role of social media and online communities cannot be overstated in facilitating these global connections. Platforms like Instagram and Etsy have become showcases for artisanal products, while forums and groups provide spaces for discussion, support, and collaboration. Through these digital channels, craftsmen find not just colleagues and partners but often lifelong friends, bridging distances with shared passions.

In essence, global collaborations in craftsmanship are a testament to the power of unity. They reflect an age where the collective effort and mutual respect for diverse traditions and techniques can lead to innovations that no single individual could achieve alone. It's a vibrant ecosystem of exchange, learning, and mutual growth that enriches not just the craftsmen involved but also the global community that gets to experience their creations.

The beauty of these collaborations lies not just in the final product but in the journey itself. The stories of cross-cultural partnerships, the blending of ancient techniques with modern innovations, and the mutual respect and learning are what truly define this era of global craftsmanship. They inspire not just craftsmen but also society at large, highlighting the importance of collaboration, diversity, and innovation.

Looking ahead, the potential for global collaborations in craftsmanship is boundless. As technology continues to evolve, so too will the ways in which craftsmen connect, create, and innovate. The future of craftsmanship is one that is inherently collaborative, global, and inclusive, offering exciting prospects for the next generation of craftsmen.

It's vital for educators, industry professionals, and young individuals to embrace and support these global collaborations. By doing so, they not only nurture the current generation of craftsmen but also lay the groundwork for future artisans. Encouraging cross-border partnerships, facilitating international exchanges, and fostering a global community of craftsmen are pivotal steps towards a future where craftsmanship knows no borders.

In conclusion, global collaborations in craftsmanship represent a confluence of tradition and innovation on an unprecedented scale. They are the key to unlocking a future where craftsmanship is celebrated not just as a means of preserving heritage but as a vital, vibrant, and integral part of the global creative economy. As we look to the future, let us champion these collaborations, for they hold the promise of crafting a world that is more connected, sustainable, and beautiful for generations to come.

Chapter 19: Intellectual Property Issues in the Digital Age

In an era where the fusion of digital innovation and traditional craftsmanship is more prevalent than ever, the subject of intellectual property (IP) emerges as a critical arena for both novices and veterans in the field. As craftsmen integrate technology into their practices, from 3D printing to virtual design, the lines defining originality and ownership blur, posing new challenges and opportunities within the digital framework. This chapter delves into the complexities of protecting your designs in a world where replication can be as simple as a click of a button. Understanding copyright challenges becomes crucial, not just for legal safeguarding but for fostering a culture of respect and recognition among the global crafts community. We explore effective strategies to navigate these waters, aiming to empower artisans to embrace the digital age while upholding the integrity of their creations. It's about striking a balance between open innovation and the protection of intellectual property, ensuring that craftsmen can share their work with confidence and receive the acknowledgment they rightfully deserve. This conversation is pivotal as we forge ahead, redefining the parameters of craftsmanship in the digital age, and securing a place for traditional skills in the future landscape of creation.

Protecting Your Designs

In today's world, where craftsmanship meets the digital age, protecting your designs has become more critical than ever. The marriage of traditional skills with new technologies has opened up a world of opportunities, yet it has also presented a unique set of challenges - particularly, how to safeguard your intellectual property in a globally connected landscape. The key is understanding that your creations aren't just products; they're valuable assets that must be shielded from potential exploitation.

The first step in this journey of protection is recognising the value of what you've created. Whether it's a unique piece of jewellery, a hand-crafted piece of furniture, or a digital design, acknowledging its worth is crucial. This isn't just about the material costs or the hours put into its creation; it's about recognising the innovation, creativity, and craftsmanship that went into it. These elements combined give your work its unique value, making it imperative to protect.

Intellectual property law might seem daunting at first glance, but it's an artisan's best friend in the digital age. Copyrights, trademarks, patents, and designs each serve a different function and protect different types of work. Knowing which type of protection best suits your creation is the first step toward safeguarding it. This might mean copyrighting your designs, trademarking your brand name, or patenting a unique process you've developed.

Documentation is your ally. It's essential to keep detailed records of your creative process, including drafts, prototypes, and correspondence relating to the development of your work. This documentation can serve as evidence of your authorship and the originality of your design should you ever need to defend your work in a legal context. It's not the most exciting part of being a craftsman, but it's undeniably important.

With the global nature of the internet, your work can easily cross borders. While this is fantastic for reaching a wider audience, it also means you need to think about protecting your designs internationally. This might require registering your intellectual property in other countries or understanding international treaties that protect your work abroad. It's a complex landscape, but navigating it ensures your creations are protected worldwide.

Another critical aspect of safeguarding your designs is being aware of the platforms you use to showcase and sell your work. Whether it's an online marketplace, social media, or your website, understanding the terms of service and how they affect your intellectual property rights is crucial. Sometimes, without realising, you might be granting these platforms extensive rights to use your work in ways you hadn't intended.

Technology isn't just a tool for creation; it's also a means of protection. Digital watermarking, blockchain, and other technologies offer new ways to secure your intellectual property. Utilising these tools can help prove ownership, timestamp your creations, and even track and manage how your designs are used across the digital landscape.

Education is pivotal. The more you know about intellectual property rights, the better equipped you'll be to protect your work. This doesn't mean you have to become a legal expert, but having a basic understanding of your rights and the measures available to enforce them can make all the difference. There are plenty of resources available designed to help creators navigate these waters.

Don't overlook the power of community. In the world of craftsmanship, there's strength in numbers. Connecting with other creators, sharing experiences, and offering support can provide valuable insights into protecting your work. Moreover, there are industry groups and associations that campaign for better protection of intellectual property and offer resources to their members.

Enforcing your rights is sometimes necessary. Despite your best efforts, there may come a time when you need to take action against infringement. This could mean sending a cease and desist letter, negotiating a settlement, or, in some cases, pursuing legal action. It's not an easy process, but defending your work is crucial for maintaining its value and your integrity as a creator.

Licensing your work can also be an effective way to protect it. By setting clear terms and conditions through a licensing agreement, you maintain control over how your work is used. This can provide a steady income stream while ensuring your designs are used in ways that align with your values and brand.

It's worth considering joining forces with other creators for collective protection strategies. Group registration of designs, collective licensing agreements, or pooled resources for legal protection can offer a more robust defence against infringement. This cooperative approach can be particularly beneficial for independent artisans and small-scale creators.

Remember, protecting your designs isn't just about safeguarding them against misuse; it's about ensuring your creations can continue to have a positive impact. As craftspeople, our work has the power to inspire, provoke thought, and bring beauty into the world. By protecting our designs, we ensure this legacy can endure.

To sum up, while the challenges of protecting your designs in the digital age are real, they're not insurmountable. With the right knowledge, tools, and support, you can safeguard your creations and continue to thrive as a craftsman in this new era. Your work is a reflection of your talent, dedication, and innovation. It deserves to be protected.

As we move forward, embracing the fusion of tradition and technology, let's also embrace the responsibility to protect the very heart of what makes our crafts unique. Let this journey of protection be a source of strength, driving the evolution of craftsmanship into the future.

Copyright Challenges

In the evolving landscape of craftsmanship, where traditional methods meet the cutting-edge capabilities of digital technology, we stand at an intersection fraught with challenges and opportunities. Among these, copyright issues represent a formidable challenge for modern craftsmen navigating the digital age. This section delves into the complexities of copyright in today's world, where the preservation of originality and innovation often clashes with the ever-expanding realms of accessibility and sharing.

The foundation of copyright law was established in a world vastly different from our own, designed to protect the rights of creators and encourage the production of creative works. However, as we march further into the digital era, the original ideals of copyright law are continually tested. Craftsmen, artists, and creators find themselves grappling with a landscape where their works can be replicated, modified, and distributed globally with a few clicks.

The ubiquitous nature of the Internet has led to an explosion of information sharing and collaboration, fueling incredible innovation and creativity. Yet, this digital proliferation also poses a significant risk to the originality and uniqueness of craftwork. The ease with which digital files can be copied and shared has created an environment where the lines between inspiration, imitation, and infringement blur more than ever before.

For craftsmen embracing digital tools and platforms, copyright serves both as a shield and a sword. It provides a legal framework to protect their creations from unauthorized use, yet it also imposes limitations on how these creations can be shared and explored by others. In a world that values open-source knowledge and communal innovation, striking a balance between protection and freedom becomes a delicate dance.

Consider the rise of 3D printing technology, which allows for the physical replication of objects from digital models. This advancement presents a fascinating conundrum: how does one protect the copyright of a physical object that can be easily scanned and reproduced? The legal precedents in this arena are still evolving, challenging craftsmen to navigate uncharted waters.

The phenomena of remix culture further complicates the copyright landscape. In an artistic and craft-based context, remixing involves taking existing works and reimagining them to create something new and original. While some view remixing as a form of flattery and a catalyst for creativity, others see it as a threat to the sanctity of the original work. The question then arises: at what point does inspiration or homage cross the line into copyright infringement?

For educators, industry professionals, and the young innovators of today, understanding the intricacies of copyright law is essential. It is not merely about protecting one's work but also about fostering an environment where creativity can flourish responsibly. This entails a continuous learning process, keeping abreast of changing laws and adapting to the new norms of a digital-first world.

Therein lies an opportunity for dialogue and advocacy. By voicing their concerns and engaging with policymakers, craftsmen can play a pivotal role in shaping copyright laws that reflect the realities of modern creation and distribution. Collaboration with legal experts, industry peers, and digital platforms can pave the way for solutions that uphold both the rights of creators and the ethos of open innovation.

Moreover, the advent of blockchain technology offers a novel approach to copyright management. By enabling the secure, transparent tracking of works and transactions, blockchain presents a possible future where copyright infringement is more easily detected and managed. This digital ledger system could revolutionize how creative works are copyrighted, licensed, and monetized, providing an additional layer of protection for digital artisans.

It's important for craftsmen to approach copyright not as a barrier but as an aspect of their trade to be understood and navigated with care. Knowledge and vigilance are key in ensuring that one's work is protected while also respecting the copyright of others. This includes seeking proper licensing for materials used, understanding fair use, and recognising the difference between copyright and other forms of intellectual property rights.

As we forge ahead, the spirit of craftsmanship must embrace both innovation and respect for intellectual property. The future of the craft lies in the hands of those who can skillfully maneuver through the complexities of the digital age, protecting their legacy while contributing to a culture of shared knowledge and mutual respect.

Ultimately, the challenge of copyright in the digital age invites us to reflect on the meaning of creativity and ownership. In a world where ideas can traverse the globe in an instant, the value of originality and personal expression takes on new weight. For the craftsmen of today and tomorrow, navigating copyright issues is not just a legal imperative but a core component of their craft's integrity and sustainability.

As we continue to push the boundaries of what it means to be a craftsman in the digital age, let us do so with a mindful approach to copyright. Let's engage with the challenges it presents not as obstacles but as opportunities to reaffirm our commitment to innovation, respect for creativity, and the pursuit of a craft that stands the test of time. The path forward is one of balance, understanding, and continuous adaptation, ensuring that craftsmanship remains a vibrant and valued part of our cultural fabric.

In conclusion, the digital age demands a new breed of craftsman: one who is not only skilled in their trade but also savvy in the rights that protect it. As we embrace the technological tools at our disposal, let's also fortify our knowledge of copyright, ensuring that our creations are not just seen and appreciated but also protected in this ever-evolving digital landscape.

Chapter 20: The Role of Festivals and Fairs in Promoting Craftsmanship

In the tapestry of modern society, festivals and fairs stand out as vibrant threads that weave together the past and present of craftsmanship, offering a rich palette for both seasoned artisans and those new to the trade to display their skills and innovations. These gatherings are not merely events; they are the crucible in which traditional crafts find new life and where the fusion of technology and timeless techniques can be showcased to an appreciative audience. Beyond the surface spectacle, festivals and fairs serve as critical networking hubs, spaces where connections are forged between craftspeople, enthusiasts, and industry insiders, opening doors to collaboration and growth. They provide a unique platform for local talent to gain visibility, often propelling small-scale artisans into the broader market. The spirit of a fair or festival, with its communal appreciation of craft, ignites a collective recognition of beauty, skill, and innovation, fostering an environment where tradition is respected and the new is celebrated. As we delve into the role these gatherings play, it's clear they are not just celebrations of craft but pivotal moments that shape the direction of craftsmanship, encouraging a new generation to blend the wisdom of the past with the possibilities of the future.

Showcasing Local Talent

In the vibrant world of festivals and fairs, there lies an unparalleled opportunity for local artisans and craftsmen to step into the limelight, bringing forth their unique abilities and creations. These events serve not merely as platforms but as catalysts for the appreciation and recognition of local talent. They're the stages where tradition meets innovation, and where local heroes find their voices amplified amidst the bustling crowds.

The essence of local craftsmanship, with its rich textures and intricate designs, finds a home in the hearts of festival-goers, many of whom are on a quest for something genuine, something that tells a story. It's at these gatherings that craftsmen don't just sell products; they share pieces of their journey, their heritage, and their passion. This direct engagement with the audience not only enhances the customer experience but also builds a loyal community around the craft.

For young artisans, especially those navigating the crossroads of tradition and modernity, festivals and fairs present an invaluable learning ground. They're spaces of live feedback and interaction, where the reactions of the audience—a smile, a nod, an intrigued question—can offer insights far beyond what digital platforms can provide. In essence, they're the physical touchpoints in an increasingly virtual world.

Moreover, festivals champion the cause of sustainability in craftsmanship. By showcasing local talent, they encourage the patronage of local productions, which often have a smaller carbon footprint compared to mass-produced goods. This aligns seamlessly with the growing environmental consciousness among consumers, particularly millennials and Generation Z, who are more inclined towards brands and products that advocate for sustainable practices.

These events also act as incubators of innovation. They provide a space for artisans to experiment with new materials, designs, and techniques before a live audience. The immediate interaction and feedback can spark new ideas or lead to the refinement of existing ones, thereby driving the evolution of crafts in real-time.

The storytelling aspect of showcasing local talent at festivals cannot be overstated. Each craft has a story—of its origin, its evolution, and its relevance in today's world. Festivals give these stories a voice, allowing craftsmen to connect with their audience on an emotional level, fostering a deeper appreciation and understanding of their work.

Aside from fostering community and sustainability, these events significantly contribute to the local economy. They provide craftsmen with direct access to markets, reducing the dependency on middlemen and enhancing their earnings potential. This economic upliftment can be particularly transformative in rural or underrepresented communities, where traditional crafts are often undervalued.

Networking is another key advantage. Artisans and craftsmen find themselves in a hive of potential collaborators, mentors, and patrons. It's not uncommon for partnerships to be formed, projects to be conceptualised, and skills to be exchanged in these lively gatherings. The spirit of camaraderie that pervades these events is potent, emboldening even the most reticent of craftsmen to reach out and engage.

The educational value of festivals and fairs should also be highlighted. They serve as open-air classrooms where the curious mind can learn about different cultures, techniques, and histories through the lens of craftsmanship. For educators and students alike, these events offer rich, tangible resources that complement academic learning.

But perhaps the most significant aspect of showcasing local talent is the empowerment it brings. For many artisans, gaining recognition and appreciation for their work is deeply affirming. It validates their efforts, encourages persistence, and inspires confidence. In turn, this empowerment can spur innovation, leading to the development of new products, designs, and methods.

In conclusion, festivals and fairs are much more than mere events; they're essential conduits for cultural exchange, economic development, and communal growth. By showcasing local talent, they not only preserve and promote traditional crafts but also pave the way for a future where craftsmanship continues to thrive, evolve, and astonish. As we navigate the intricate dance between preserving our heritage and embracing innovation, these gatherings remind us of the beauty in creation, the strength in community, and the enduring power of craftsmanship.

As we look to the future, it's clear that the role of festivals and fairs in showcasing local talent will only grow in importance. They are the breeding grounds for the next generation of craftsmen and women, offering a platform that is both traditional and innovative. In embracing these events, we foster an ecosystem where craftsmanship can flourish, adapting to the changing times while staying true to its roots.

Let us then continue to support, participate in, and celebrate these festivals and fairs, for in doing so, we not only champion the talents of today but also nurture the legacy of craftsmanship for future generations. The story of local craftsmanship is one that deserves to be told, retold, and celebrated, now more than ever.

Networking Opportunities

In the realm of craftsmanship, the significance of festivals and fairs cannot be overstated, particularly when it comes to the invaluable networking opportunities they afford. These gatherings are not merely events; they're vibrant ecosystems where ideas proliferate, collaborations begin, and future industry leaders connect. At the heart of every fair and festival lies the potential for craftsmen to broaden their horizons, both personally and professionally.

Festivals and fairs offer a unique platform for craftsmen to showcase their talent and work amongst their peers, potential collaborators, and industry veterans. It's a place where one's craft does not just stand as a testament to individual skill but acts as a beacon, attracting like-minded individuals and potential business partners. The environment fosters a sense of community and mutual respect that is hard to find in other professional settings.

For young craftsmen, particularly those from the millennial and Gen Z generations, these events can serve as a gateway to integrating traditional skills with modern technological innovations. Engaging in conversations, attending workshops, and participating in panel discussions can provide unprecedented access to new ideas and technologies that could elevate their craft to new heights.

These gatherings also provide a fertile ground for mentorship opportunities. Seasoned professionals often attend festivals and fairs with the intention of passing on knowledge and may be on the lookout for enthusiastic individuals keen on learning the trade. Engaging with experienced craftsmen can offer insights into not just the technicalities of the trade but also the business aspects of running a successful craft-based venture.

Networking at these events can lead to collaboration opportunities that might not have been possible otherwise. Whether it's a joint project between different craftsmen or a partnership with suppliers of raw materials, fairs and festivals can kickstart symbiotic relationships that benefit all parties involved. Collaborations such as these are crucial for innovation and can lead to new products or methodologies that push the boundaries of traditional craftsmanship.

Furthermore, participating in or attending these events exposes craftsmen to a larger audience, including potential customers and clients. It's an effective way to build one's brand and market one's craft directly to interested parties. Engaging face-to-face with potential buyers offers immediate feedback and can forge emotional connections that digital marketing campaigns might struggle to achieve.

It's also an arena for craftsmen to learn about the competition, stay abreast of industry trends, and understand market demand. Observing and interacting with fellow craftsmen allows for a direct comparison of one's own work with that of others, providing valuable insights into areas of improvement and innovation.

Beyond professional growth, festivals and fairs offer emotional support to craftsmen. These events can be a powerful reminder that one is part of a larger community of makers and artisans, all sharing similar challenges and triumphs. Such connections can be profoundly motivational, particularly when facing the everyday hurdles of the craft.

Another vital aspect of these gatherings is the opportunity to engage with a global audience and collaborators. With the world becoming increasingly connected, festivals and fairs often attract international visitors and exhibitors. Such diversity can lead to cross-cultural exchanges, opening up new markets and inspirations for craftsmen.

For those looking to bridge the gap between traditional craftsmanship and the digital world, these events can serve as a platform to explore digital marketing techniques and e-commerce opportunities. Meeting with digital-savvy marketers and fellow craftsmen who have successfully navigated the transition to online sales can provide valuable lessons and strategies.

Festivals and fairs also act as a microcosm for the larger economic and social impact of craftsmanship. Through meetings, debates, and informal conversations, craftsmen can engage with policy-makers, advocates, and the media, raising awareness about the importance of their work and voicing their needs and challenges.

Moreover, the atmosphere of creativity and innovation that pervades these events is infectious. Being surrounded by the best in the field, witnessing their creations, and engaging in creative discussions can ignite new passions and reinforce the love for one's own craft.

The potential for these networking opportunities to lead to lifelong friendships and collaborations should not be underestimated. Often, the connections made during these events go beyond professional networking, leading to personal growth and an expanded worldview.

In conclusion, festivals and fairs are much more than an opportunity to sell and showcase craftsmanship. They represent a fertile ground for networking, learning, and inspiration. For craftsmen at any stage of their career, these events provide a pathway to not just succeed in their craft, but to thrive within a community of like-minded individuals, forging a future where tradition and innovation coexist harmoniously.

As we move forward, embracing the integration of technology and innovation with traditional craftsmanship, these networking opportunities will play a pivotal role in shaping not just individual careers but the future of craftsmanship as a whole. It is here, amidst the hustle and bustle of festivals and fairs, that the seeds of tomorrow's artisanal landscape are sown.

Chapter 21:
The Power of Storytelling in Craftsmanship

In a world where the palpable texture of handcrafted objects tells tales of dedication and innovation, storytelling emerges as an indispensable tool, interweaving the past, present, and future of craftsmanship. *The Power of Storytelling in Craftsmanship* underscores how narratives not only connect us to the essence of the craft but also serve as a pivotal mechanism for branding and forming an identity in the digital age. Through stories, craftsmen can convey the meticulous labour, the inherited techniques refined through generations, and the bursts of creativity that animate each creation. These narratives go beyond mere marketing; they offer a bridge for consumers to experience the soul behind the object, fostering a deeper appreciation and a connection that transcends the physical product. For educators, industry professionals, and the youth enamoured by the blend of tradition and innovation, mastering the art of storytelling is not just about selling a product but about preserving heritage, inspiring future artisans, and embedding personal and collective identities within creations. Hence, as we navigate the evolving landscape of craftsmanship, embracing storytelling becomes not merely an option but a necessity, uplifting the value of handmade goods in a world dominated by mass production and fostering a future where craftsmanship flourishes, rich in stories that resonate across time.

Connecting Through Narratives

In the vast and vibrant world of craftsmanship, stories hold a pivotal role. They are the threads that weave the very fabric of our cultural heritage, intertwining generations, techniques, and aspirations. In this section, we delve into the significance of narratives in building a profound connection between craftsmen and their audience, a connection that transcends the mere utility of crafted objects and embraces the rich tapestry of human experience.

At the heart of every crafted piece lies a narrative - a story of imagination, perseverance, and intricate skill. These stories not only provide a glimpse into the creator's soul but also offer a bridge to the past, linking modern craftsmanship with ancient traditions. Through narratives, craftsmen can communicate their passion and dedication, making their creations resonate with those who seek more than just the superficial.

Narratives serve as a powerful tool for connecting with the younger generations, particularly Millennials and Gen Z. Raised in the digital era, these individuals are often seeking authentic experiences and meaningful engagement. Stories of craftsmanship, with their emphasis on sustainability, individuality, and creativity, naturally appeal to these values, creating a sense of belonging and identification among a demographic that yearns to make a difference.

Moreover, narratives have the unique capability to demystify the creative process. They unveil the behind-the-scenes struggles, the moments of inspiration, and the meticulous effort that goes into every masterpiece. This transparency not only fosters appreciation and respect for the craft but also inspires others to embark on their own creative journeys.

Education and skill development, too, can greatly benefit from the integration of narratives. Traditional crafts, when narrated as living stories, become appealing subjects of study, bridging the gap between theoretical knowledge and practical application. Through storytelling, educators can ignite a passion for learning among students, motivating them to delve deeper into the realm of craftsmanship.

Digital platforms have emerged as powerful mediums for sharing these narratives. Social media, blogs, and online forums allow craftsmen to share their journeys, challenges, and achievements with a global audience. These platforms not only help in building a community but also in establishing a personal brand, which is essential in the contemporary market.

Yet, the power of narratives extends beyond marketing and branding. They imbue crafted objects with a soul, transforming them from mere commodities into symbols of human endeavour and creativity. A narrative-driven approach in craftsmanship encourages consumers to look beyond the price tag and appreciate the intrinsic value of handcrafted goods. This shift in perception is crucial for promoting sustainable practices and supporting artisan communities.

In a world where mass production and consumerism prevail, narratives remind us of the significance of quality over quantity. They encourage a mindful approach to consumption, where each purchase is viewed as an investment in creativity, hard work, and cultural preservation. By choosing crafts with stories, consumers become active participants in keeping traditional skills alive and relevant.

Narratives also play a crucial role in fostering cross-cultural exchanges. In the global village of the 21st century, craftsmen have the opportunity to share their heritage with the world and, in turn, learn from the diverse traditions of others. These exchanges not only enrich the craft itself but also promote empathy, understanding, and unity among cultures.

Furthermore, the act of storytelling is therapeutic. For many craftsmen, narrating their journey is a way of reflecting on their growth, challenges, and achievements. It provides a sense of purpose and fulfilment, contributing to their mental well-being. In a similar vein, listeners find inspiration and solace in these stories, often seeing reflections of their own aspirations and struggles.

Looking ahead, the integration of technology in storytelling opens new realms of possibilities. Augmented reality, virtual reality, and interactive platforms can provide immersive experiences, bringing narratives to life in ways previously unimaginable. Innovators in the field of craftsmanship have the opportunity to pioneer these technologies, creating captivating storytelling experiences that engage and inspire.

Despite the countless benefits, effectively connecting through narratives requires skill and authenticity. The story must be genuine and told with heartfelt emotion; otherwise, it risks being dismissed as mere marketing fluff. Craftsmen must find their unique voice, one that reflects their values, vision, and identity.

In conclusion, narratives are more than just stories; they are the essence of craftsmanship in the modern world. They build connections, inspire change, and imbue crafted objects with meaning. As we move forward, embracing the power of storytelling will be instrumental in shaping the future of craftsmanship, making it relevant, valued, and impactful in the digital age.

Let us all, then, become storytellers in our own right, using narratives to bridge gaps, to educate, to inspire, and to connect. For in the end, it is through our stories that we leave a legacy, one that resonates with the future while honouring the past.

Branding and Identity

In today's digitally connected world, the power of storytelling in craftsmanship extends far beyond the creation of a product. It's about building a unique brand and identity that resonates with your audience. This chapter will delve into how craftsmen can harness the narrative of their art to form a lasting impression on both their industry and their patrons. Understanding how to effectively communicate your brand's story is paramount in setting you apart in a saturated market.

Firstly, the essence of a brand lies not just in its logo or name but in every interaction you have with your audience. For craftsmen, this means that each piece of work carries with it a story—of tradition, of skill, or of innovation. It's these stories that imbue your products with a personality, a character that people can connect with. When customers buy from you, they're not just acquiring an item; they're becoming part of the narrative.

Creating a memorable brand identity as a craftsman requires a deep understanding of your own story. Why did you choose this path? What inspires your designs? How do your methods differentiate you from others in your field? Answering these questions through your marketing materials, social media presence, and direct interactions will help form a coherent narrative that speaks to your target audience.

Moreover, the impact of visual storytelling cannot be overstressed. In a realm where Instagram, Pinterest, and other visual platforms reign supreme, the ability of high-quality images and videos to convey the intricacies of your work is unparalleled. These visuals are not mere advertisements; they are chapters in your ongoing saga, inviting viewers to step into your world.

Collaboration also plays a critical role in branding and identity. By aligning with artists, designers, and influencers who share your values and aesthetic, you can create compelling stories that capture the essence of your brand. These partnerships not only broaden your reach but also add depth to your brand's narrative, illustrating its relevance and application in various contexts.

Emphasising the sustainability and ethical considerations of your practice within your brand's storytelling appeals to today's environmentally conscious consumer. Highlighting how your craftsmanship contributes to a more sustainable and socially responsible future can be a powerful differentiator in the marketplace.

It's equally important to ensure that the narrative is authentic. In an age where consumers are bombarded with marketing from every direction, sincerity stands out. Your audience can sense when a brand is genuine in its storytelling, and this authenticity fosters stronger, more meaningful connections to your brand.

Consider the role of customer feedback and user-generated content in shaping your brand's narrative. Encouraging your customers to share their experiences with your products can provide invaluable insights into how your brand is perceived and can further amplify your story through the lens of those who have directly engaged with your craft.

Furthermore, the digital landscape offers a myriad of tools for tracking and analysing the impact of your brand's story. Leveraging these tools can help you refine your messaging, understand your audience better, and tailor your story to resonate more deeply with them.

In addition, don't overlook the power of local identity in crafting your brand's narrative. Embracing the traditions, materials, or motifs of your region can add a layer of depth and authenticity to your story that can appeal to both local and global audiences.

Another aspect to consider is how you communicate your brand's evolution. Your identity isn't static but something that grows and changes with time. Sharing this journey with your audience can be an engaging way to keep them involved and invested in your brand.

In the crafting industry, it's also essential to be mindful of how your brand's narrative integrates with the broader community. Participating in local events, workshops, or collaborations can embed your brand into the fabric of the crafting community, further solidifying your identity and presence.

Lastly, the end goal of establishing a strong brand identity is not just to sell a product but to create an experience that your customers value. This experience is what ultimately turns one-time buyers into lifelong fans and advocates for your brand.

In conclusion, the intersection of craftsmanship, branding, and storytelling offers a potent blend for creating a vibrant, enduring brand identity. By weaving the rich tapestry of your craft's story into every facet of your brand, you not only elevate your art but also forge a deeper connection with your audience. It's this connection that transforms your craft from a simple trade into a legacy.

As we progress into an era where the lines between traditional trades and modern innovation continue to blur, the stories we tell through our crafts will play an increasingly significant role in defining the future landscape of craftsmanship. It's through these narratives that a new generation of craftsmen will inspire, innovate, and ultimately, leave their mark on the world.

Chapter 22: Addressing Climate Change through Craftsmanship

In the vital pursuit of combatting climate change, the role of craftsmanship—an often overlooked ally—emerges with compelling clarity. This chapter delves into how the inherited wisdom of traditional crafts, combined with revolutionary green technologies, paves a sustainable future that not only respects the earth but also enriches our cultural heritage. The marriage of age-old techniques with innovative methods such as the circular economy and sustainable production practices offers a blueprint for creating products that not only stand the test of time but also minimise environmental impact. As craftsmen around the globe embrace materials and processes that reduce waste and energy consumption, they set a benchmark in the fight against climate change. This narrative isn't just about mitigating harm; it's a compelling vision of craftsmen at the forefront, wielding their skills as both a shield and a spear in the battle for a better planet. Through this exploration, we uncover not just the possibility, but the necessity of infusing our trades with the principles of sustainability, demonstrating that each stroke of the chisel, each turn of the potter's wheel, and every stitch in fabric can indeed contribute to the monumental task of making our world a healthier, more harmonious place.

Sustainable Production Methods

In the journey towards addressing climate change, the traditional craftsmanship sector is uniquely positioned to lead by example. Sustainable production methods are not just a trend or a marketing tool; they represent a fundamental shift in how we conceive the making, distribution, and consumption of goods. This chapter delves into the transformative potential of integrating sustainability into the heart of craftsmanship, exploring both the challenges and the immense opportunities that lie ahead.

Sustainability in craftsmanship begins with the source of materials. The choice to use eco-friendly or recycled materials is a significant step towards reducing the environmental impact of production. This approach not only lessens the depletion of natural resources but also sets a standard for responsible sourcing in industries far beyond craftsmanship. The art of selecting materials with care and consciousness becomes a testament to a craftsman's commitment to the health of our planet.

Water usage and waste management are critical areas where sustainable practices can make a profound difference. Traditional methods of production often overlook the ecological cost of excessive water consumption and waste generation. Modern craftsmen are innovatively addressing these issues by implementing systems for water recycling and waste minimization in their workshops, demonstrating that efficiency and environmental stewardship can go hand in hand.

Energy consumption is another aspect where sustainability can be deeply embedded into craftsmanship. By utilising renewable energy sources such as solar or wind power, workshops can drastically reduce their carbon footprint. The transition to green energy is not only beneficial for the environment but can also lead to economic savings in the long run, showcasing the dual advantages of sustainable practices.

The concept of the circular economy holds immense promise for the future of craftsmanship. This model emphasizes the need to design products with their end-of-life in mind, focusing on durability, repairability, and the possibility of recycling or upcycling. Craftsmen who embrace these principles are paving the way for a world where goods are valued not just for their immediate use but for their potential to be repurposed and reused.

Another sustainable production method involves the adoption of digital technologies. Techniques such as 3D printing in artisanship not only allow for precision and customization but also significantly reduce waste by using only the necessary amount of material. Furthermore, digital tools enable craftsmen to experiment and prototype with minimal environmental impact, merging innovation with sustainability.

Transportation and packaging also provide areas for sustainable improvements. By choosing local materials and markets, craftsmen can reduce the carbon emissions associated with long-distance shipping. Additionally, eco-friendly packaging made from recycled materials or no packaging at all highlights a commitment to reducing single-use plastics and waste.

Consumer education is a powerful tool in the sustainable craftsmanship movement. By informing customers about the ecological benefits of their purchases, craftsmen can inspire wider community engagement with sustainability. This dialogue between maker and buyer fosters a culture of responsible consumption, where every purchase is considered in light of its environmental impact.

Collaboration among craftsmen also amplifies the impact of sustainable practices. Through sharing resources, knowledge, and innovations, the crafting community can collectively advance towards more environmentally friendly production methods. These partnerships, whether local or international, are instrumental in creating a unified front against climate change.

The challenges in transitioning to sustainable production methods cannot be understated. Initial investments, learning new techniques, and finding eco-friendly materials can be daunting tasks. However, the long-term benefits for the planet, combined with an increasing consumer demand for sustainable products, create a compelling case for embracing this change.

Public policy and industry regulations also play a crucial role in supporting sustainable craftsmanship. Advocacy for incentives, subsidies, and laws favouring eco-friendly production can help level the playing field, making it easier for craftsmen to adopt sustainable practices. This governmental support is crucial for the widespread adoption of sustainability in the industry.

The impact of sustainable production on the mental health and well-being of craftsmen is an area of emerging interest. Working in an environmentally conscious manner fosters a sense of purpose and connection with the planet, contributing to overall job satisfaction and well-being. This emotional and psychological benefit enriches the crafting experience, making sustainability not just a professional choice but a personal one as well.

Looking ahead, the possibilities for sustainable craftsmanship are limitless. Innovations in eco-friendly materials, green technologies, and circular economy models are constantly emerging, offering new opportunities for craftsmen to refine their practices. The future of craftsmanship is undoubtedly sustainable, combining respect for the environment with the enduring value of handcrafted goods.

As we forge ahead in the fight against climate change, the role of craftsmanship and sustainable production methods cannot be overstated. It is time for craftsmen to lead by example, proving that it is indeed possible to honour our heritage while safeguarding the future of our planet. By embracing sustainability, craftsmen can contribute to a more resilient, equitable, and environmentally conscious world.

In conclusion, the intersection of craftsmanship and sustainability represents a beacon of hope in the context of climate change. Through thoughtful innovation, mindful material selection, and a commitment to green practices, modern craftsmen are demonstrating that it is possible to create beautifully crafted goods in harmony with the earth. The path towards a sustainable future in craftsmanship is both an inspiring challenge and a profound opportunity to make a lasting difference.

The Circular Economy

The conversation around tackling climate change is evolving, and at the heart of this shift is the concept of the circular economy. It's a transformative idea that touches on sustainability, but goes further, challenging us to rethink our entire approach to production and consumption. In the realm of craftsmanship, this principle doesn't just resonate—it heralds a new era of innovative thinking and creative solutions.

At its core, the circular economy aims to design out waste. It's about creating systems where everything has value and nothing is wasted, in stark contrast to the traditional linear model of 'take, make, dispose.' For craftspeople, this notion isn't just inspirational; it's a practical blueprint for sustainable practice. It impels artisans to consider the lifecycle of their creations, from sourcing materials to the end-of-life phase, ensuring that each stage minimises environmental impact.

Incorporating the circular economy into craftsmanship means looking back to look forward. Historically, artisans were the original recyclers—materials were scarce and valued, so nothing was wasted. Today's craftsmen can draw on this heritage, applying age-old principles with new technologies and materials to innovate sustainably.

Imagine, for example, a woodworker who sources timber from sustainable forests or uses offcuts from local industries, minimising waste. Or a metalworker who melts down scrap metal to forge new pieces. This is the circular economy in action—materials circulating within the economy for as long as possible, adding value and reducing environmental impact.

Furthermore, the circular economy encourages us to reimagine waste itself. Waste material from one craft process could become the raw material for another, fostering a culture of collaboration among craftspeople. This not only enhances sustainability but also opens up new avenues for creativity and innovation.

Adopting circular principles also presents a challenge to the way we perceive craftsmanship and its products. In a world accustomed to mass production and disposable goods, craftsman-made items stand out not just for their quality, but for their sustainability. Each piece tells a story of thoughtful design, meticulous selection of materials, and a deep respect for the environment.

For educators and industry professionals, the circular economy offers a framework for teaching and practice. It's an opportunity to instil in emerging craftsmen the values of sustainability, resourcefulness, and responsibility. This isn't just about protecting the environment—it's about crafting a future where trade and tradition contribute actively to the wellbeing of our planet.

Digital technology plays a pivotal role in this transition. Innovations like 3D printing in artisanship, which allows for precise material use and minimal waste, are a testament to how technology can support the circular economy. IoT devices can track the use and wear of tools, optimising their lifetime and reducing the need for replacements. This fusion of the traditional with the digital opens up new possibilities for sustainable craftsmanship.

Embracing the circular economy also means changing how products are designed. Craftspeople can lead the way in creating products that are made to last, can be easily repaired, or are modular, allowing parts to be replaced or upgraded without discarding the whole item. This approach not only reduces waste but also strengthens the bond between the maker and the consumer, as each piece is designed with care, purpose, and an eye towards longevity.

The benefits of the circular economy extend beyond the environment. For craftsmen, it's a way to differentiate themselves in a crowded market, appealing to consumers who increasingly value sustainability. It's also an opportunity to reduce costs by using materials more efficiently and tapping into new markets for recycled and upcycled products.

However, transitioning to a circular model is not without its challenges. It requires a shift in mindset, not just among craftsmen but also consumers, policy-makers, and the wider industry. It calls for collaboration across sectors and disciplines, breaking down silos to create a shared vision for a sustainable future.

Educational institutions have a crucial role to play in this transition. By integrating the principles of the circular economy into curricula, they can prepare the next generation of craftspeople to think differently about materials, design, and waste. This education is not just technical; it's also ethical, focusing on the broader implications of our making choices.

For young individuals curious about crafting, the circular economy offers a vision of a career that is not only creatively fulfilling but also environmentally responsible. It's an invitation to innovate, to challenge the status quo, and to make a tangible difference in the world.

In conclusion, the circular economy represents a profound shift in how we approach craftsmanship in the context of climate change. It's a fusion of old and new—a celebration of tradition, innovation, and sustainability. As we look towards a future where the health of our planet is paramount, the principles of the circular economy offer not just hope, but a clear path forward. By embracing these principles, craftspeople can lead the way in creating a world where beauty, functionality, and environmental stewardship are intrinsically linked.

Chapter 23: Accessibility and Inclusion in the World of Crafts

In a world where the spirit of creativity knows no bounds, the importance of fostering an environment of accessibility and inclusion within the realm of crafts cannot be overstated. It's here, in the meticulous intertwining of hands and materials, that we're reminded of the power embedded in ensuring everyone has the opportunity to contribute to and partake in the cultural and artistic expressions that crafts provide. This chapter delves into the heart of breaking down barriers that have historically marginalised certain groups from fully participating in craft-making and -appreciating activities. It's about levelling the playing field so that regardless of one's physical ability, socioeconomic background, or any other factor that might have served as a hurdle, there's a welcoming space at the crafting table. By championing inclusive practices, we're not only enriching the craft community but also embracing a diversification of perspectives that inherently enhances creativity and innovation. Through shedding light on successful initiatives and providing actionable guidelines, this chapter aims to inspire readers to actively contribute towards an inclusive crafting culture. A culture where the only prerequisite is a passion for craftsmanship, and where the unique stories and backgrounds of every individual are considered invaluable assets that enrich the collective tapestry of the modern crafting world.

Breaking down Barriers

The world of crafts is evolving, becoming a vibrant tapestry woven from countless diverse threads, each representing different skills, backgrounds, values, and visions for the future. However, this rich tapestry can only achieve its full potential if we actively work to dismantle the barriers that prevent full participation and inclusion in this dynamic field. From physical to digital, economic to cultural, the barriers facing aspiring craftsmen today are as varied as the crafts themselves.

Firstly, physical accessibility remains a significant challenge. Traditionally, craftwork necessitates hands-on interaction, posing an obstacle to those with physical disabilities. Yet, through technology, we're witnessing an era where adaptive tools and methods are emerging, allowing craft to be more accessible than ever. Innovations such as voice-activated software and customisable workstations are not just tools; they are lifelines to creativity and vocational fulfilment for individuals who might otherwise be excluded.

Economically, the costs associated with training, materials, and necessary equipment can be prohibitively high. This financial barrier often discourages talent from underprivileged backgrounds from pursuing a career in crafts. Scholarships, grants, and community-supported workshops are essential in leveling the playing field, ensuring that passion and talent are the only prerequisites for success in the craft world.

The digital divide further compounds the issue of economic accessibility. In our rapidly digitising world, being competent in digital tools and platforms is increasingly essential in the realm of crafts, from design and production to marketing and sales. Bridging this divide requires concerted efforts to provide digital literacy and resources to all aspiring craftsmen, regardless of their socioeconomic status.

Cultural and societal barriers also play a critical role. In many societies, traditional gender roles continue to dictate who can or should pursue certain crafts. This stereotyping not only limits individual choice but also impoverishes the craft community by narrowing the diversity of perspectives and skills. Celebrating and promoting stories of individuals who break these molds is crucial in fostering an environment of inclusion and respect.

Another significant barrier is the lack of visibility and representation within the crafts sector. Many from marginalised communities do not see themselves reflected in the world of crafts, discouraging their participation. Increasing visibility and representation through targeted outreach, mentorship programs, and inclusive showcasing can help bridge this gap.

The language of crafts, often steeped in tradition, can inadvertently become a barrier. Ensuring that the terminology, teaching methods, and promotional materials are accessible and understandable to all, including those for whom English is not a first language or who have learning difficulties, is fundamental to inclusivity.

Furthermore, there's an urgent need to reassess how we value and teach crafts at an educational level. In many educational systems, crafts are not accorded the same status as academic subjects, leading to a lack of opportunity and encouragement for young people to pursue crafts professionally. Advocating for craft education's incorporation into curriculums from an early age can cultivate appreciation and talent from the ground up.

Social media and online platforms, while powerful tools for promotion and community building, also present new barriers in terms of digital literacy and access. Educators and industry leaders must guide emerging craftsmen in navigating these spaces effectively, ensuring that digital presence enhances rather than hinders their work.

In the realm of mental health, the solitary nature of many crafts can become a barrier to well-being. Building supportive communities and networks where craftsmen can share experiences, challenges, and successes is critical. These networks can act as a safety net, promoting mental health and emotional well-being alongside professional development.

Moreover, the legal landscape of intellectual property rights presents a specialized barrier, with many craftsmen unaware of how to protect their unique creations. Education and accessible resources on copyright laws and protection mechanisms can empower craftsmen to safeguard their work and creativity against exploitation.

As we envision the future of craftsmanship, we must also consider environmental barriers. The use of non-sustainable materials and practices can not only harm our planet but also limit the long-term viability of craft practices. Embracing environmentally friendly materials and sustainable practices can thus serve as both a moral and practical guideline for the future of crafts.

The challenge of breaking down barriers in the world of crafts is not insurmountable. Through collective action, awareness, and the embrace of technology and innovation, we can build a more inclusive, diverse, and sustainable craft community. This journey requires us to not only look inward, reflecting on our practices and prejudices, but also to look forward, envisioning a future where the world of crafts is accessible to all.

Ultimately, the richness of the craft world lies in its diversity — of ideas, perspectives, materials, and methods. By committing to breaking down the barriers that limit this diversity, we can unlock a future for craftsmanship that is as boundless as our collective creativity. It's a future where everyone can not only participate but thrive and contribute to the ever-evolving tapestry of human skill and artistry. Let's forge ahead, together, into this promising and inclusive future.

Inclusive Practices

Inclusivity in the realm of crafts does not only enhance the diversity and richness of its community but also ensures that crafts remain relevant and accessible to all. By embracing inclusive practices, the world of craftsmanship can create environments where everyone feels welcomed, represented, and empowered.

At its core, inclusivity in crafts means acknowledging that diverse backgrounds can bring in fresh perspectives and innovative approaches to traditional and modern crafting techniques. It's about recognising and dismantling the barriers that might prevent individuals from underrepresented groups from participating fully in the craft community.

Accessibility is a crucial aspect of inclusivity. Making tools, materials, and learning resources available and affordable can significantly lower the barriers to entry for many aspiring craftsmen. This includes providing scholarships, grants, and free educational content online to reach individuals who might not have the financial means or physical access to brick-and-mortar schools and workshops.

Moreover, embracing digital platforms can democratise the process of learning and sharing crafts. Online forums, webinars, and tutorials offer the opportunity for people from across the globe to connect, learn, and collaborate, regardless of their geographical or physical limitations. This digital inclusion fosters a sense of global community, where everyone, from anywhere, can be both a student and a teacher.

Language, too, can be a barrier. Offering resources, tutorials, and community support in multiple languages makes the craft world more accessible to non-English speakers. This not only enriches the community with diverse cultures and traditions but also elevates the craft itself by introducing variations and innovations borne from different cultural approaches.

Sensitivity towards physical disabilities is also essential in crafting inclusivity. Workshops and online platforms should strive to provide resources and accommodations that make crafting accessible to individuals with disabilities. This could include adaptive tools, accessible workspaces, and tutorials designed with an understanding of various needs.

Representation matters immensely. Highlighting and celebrating the work of craftsmen from diverse backgrounds can inspire others to see themselves within the crafting community. It's not just about featuring diverse artists but also about including their cultural stories and perspectives, which adds depth and richness to the craft narrative.

Inclusivity also means challenging stereotypes and addressing biases within the crafting community. Workshops, discussions, and education on diversity and inclusion can help dismantle prejudices and create a more welcoming environment for everyone.

Creating mentorship opportunities for underrepresented individuals can bridge gaps between aspiring craftsmen and established professionals. Mentorship provides not just practical skills and knowledge but also the confidence and support that novices need to pursue their passions in crafts.

Events and forums should be consciously designed to be inclusive. This includes considering the location, timing, accessibility, and even the marketing of these events to ensure they are inviting to a diverse audience. It's about creating spaces where everyone feels they belong and can freely express their creativity.

Feedback mechanisms are vital in ensuring continued progress towards inclusivity. Encouraging feedback from community members about their experiences and suggestions for improvement enables organisations and leaders in the craft sector to make informed adjustments to their policies and practices.

Inclusive practices in crafts also extend to the economic aspects. Ensuring that craftsmen from all backgrounds have fair opportunities to market and sell their products can help alleviate the economic disparities within the community. This includes providing platforms and markets that are accessible to small-scale artisans and those from marginalised communities.

Finally, inclusivity in crafts is about building and nurturing a community that values and respects each member's unique contributions. It's about moving away from a competitive mindset to one of collaboration and shared growth, where the success of one is seen as the success of all.

The journey towards a fully inclusive world of crafts is ongoing and requires the collective effort of every member of the community. It's a journey that promises not just a more diverse and vibrant craft culture but also a more innovative and sustainable future for the crafting industry.

In sum, embracing inclusive practices in the world of crafts is not just a moral imperative but a strategic one. It ensures the longevity and relevance of crafts in a rapidly changing world and helps build a community that reflects the rich tapestry of human diversity. The future of crafts lies in our ability to be inclusive, accessible, and welcoming to all who wish to be a part of this creative world.

Chapter 24: The Intersection of Art and Craftsmanship

In the realm where tradition meets innovation, the blending of art and craftsmanship emerges as a fascinating dialogue enriching both fields. This convergence isn't just about applying techniques; it's a profound exploration of how the aesthetic values of art and the functional rigour of craftsmanship can coalesce to create objects and experiences that resonate on a deeper cultural and personal level. Artisans are increasingly adopting the mantle of artists, pushing the boundaries of their trades with creative flair that rivals that of their counterparts in fine arts. This evolution reflects a broader cultural shift towards valuing the unique over the mass-produced, the story behind an object as much as its utility. As we navigate the future, this chapter aims to inspire a renaissance of sorts—a call to arms for the next generation of craftsmen to envision their work as part of a larger tapestry of human expression. Through exploring the blurred lines between these two worlds, we uncover the vast potential for innovation, personal fulfilment, and cultural contribution. By embracing this intersection, we not only honour the legacy of traditional crafts but also chart a course for a future where art and craftsmanship are indistinguishably intertwined, offering a richer, more diverse canvas of human achievement.

Blurring the Lines

In the evolving tapestry of modern craftsmanship and art, the boundaries that once demarcated where art ends and craft begins have increasingly become nebulous. This shift is pivotal to understanding the transformation within the spheres of creativity and production. In an age where technology rapidly evolves and societal values lean heavily towards sustainability and innovation, the distinction between art and craftsmanship is not just fading; it's purposefully being redefined.

The age-old debate of art versus craft has often seen the former placed on a pedestal - celebrated in galleries and revered in academia, while the latter was boxed into the realm of functional, practical work. However, this traditional hierarchy is being challenged by a new generation of creators who fluidly navigate between these worlds, merging them in ways that were once unimaginable.

Consider the digital artisan, a figure emblematic of this new era. Skilled in traditional methods of creation, yet fully embracing the potential of digital tools, their works stand at the intersection of art and craftsmanship. 3D printing, once a novel technology, is now material for both practical objects and intricate art forms. Similarly, virtual reality offers not just an escape but a canvas, one where the immersive environments crafted meld technical prowess with artistic vision.

The implications of this convergence are profound. It challenges educators to rethink how craftsmanship and art are taught. Curricula that once favoured clear-cut distinctions now need to foster interdisciplinary approaches, blending skills and concepts across domains. Industry professionals, too, must adapt, recognising the value in hybrids of form and function, art and utility.

This blending of fields also reflects a broader societal shift towards valuing versatility, sustainability, and personalisation. In a world awash with mass-produced goods, the allure of the handmade, the custom, and the unique has never been stronger. The modern consumer seeks not just quality but identity in their possessions - a narrative thread that ties back to the creator and their vision. Thus, craftsmanship enriched with artistic sensibility becomes a powerful differentiator in the market.

From a sustainability perspective, this convergence encourages innovative use of materials and techniques. The artisan as artist is not just crafting objects but is often engaged in a dialogue with the environment. Whether through upcycling unexpected materials into beautiful objects or employing eco-friendly practices that challenge the status quo of production, these creators are at the forefront of a movement towards mindful consumption.

The internet and social media have acted as catalysts in this blurring of lines. Platforms that once served purely as portfolios have become stages for storytelling, where the narratives of creation are as valued as the creations themselves. This has not only democratised the field but has also fostered communities that span geographies and disciplines, encouraging collaboration and the sharing of ideas and techniques.

Moreover, the integration of artful craftsmanship into public and private spaces is redefining aesthetics in our built environment. Art installations that leverage craft techniques not only beautify spaces but also provoke thought and conversation, underscoring the role of the maker as a critical voice in societal commentary.

The re-emerging role of women in trades, breaking stereotypes and bringing fresh perspectives, also enriches this landscape. As the narratives of women artisans and their unique challenges and triumphs are woven into the broader tapestry, they add depth and complexity to the conversation around craftsmanship and art.

In harnessing the power of story, this new wave of creators makes an emphatic case for the fusion of art and craft. Through their stories, they not only capture the imagination of their audience but also inspire a new generation to explore the boundless possibilities that lie in the amalgamation of these disciplines.

The road ahead requires overcoming entrenched biases and fostering environments that celebrate and nurture the fusion of skills and ideas. Educational institutions, cultural organisations, and industry bodies all have roles to play in facilitating this shift, ensuring that the fluidity between art and craft is not just acknowledged but encouraged.

Ultimately, the blurring of lines between art and craftsmanship is more than a mere trend; it's a reflection of the evolving human need to express, connect, and belong. It speaks to a future where the labels of 'artist' and 'craftsperson' are less important than the impact and relevance of the work. As we move forward, embracing this fusion will be key to understanding not just the future of craftsmanship but of creativity itself.

In conclusion, as we navigate this intriguing confluence of art and craftsmanship, we're reminded that at the heart of both disciplines lies a common thread - the desire to create. Whether chiselled by hand or forged in the digital realm, the creations that emerge from this blend are testament to the enduring human spirit of innovation and expression. Armed with this understanding, we can look forward to a future rich with potential, crafting tomorrow with the best of today's innovations and yesterday's traditions.

As the lines continue to blur, we stand on the brink of a renaissance in making - one that champions diversity, embraces complexity, and celebrates the marriage of beauty and utility. For educators, industry professionals, and young individuals, the message is clear: the future of craftsmanship lies in its ability to evolve, adapt, and integrate. In this journey, every stitch, stroke, and code is a step towards a more vibrant, interconnected, and creative world.

Artisan as Artist

In the interwoven tapestry of modern craftsmanship, the distinction between the artisan and the artist becomes ever more nuanced. Traditionally, artisans were seen as skilled labourers, their hands weaving, carving, and constructing the necessities of daily life. Artists, on the other hand, were the visionaries, tasked with the creation of works meant to move, provoke, and encapsulate the essence of human emotion and experience. Yet, this demarcation blurs as we step into an era where craft and creativity intertwine indistinguishably. It's here, in this liminal space, that we find the artisan evolving, embodying the soul of an artist.

The narrative that positions craftsmanship and artistry on opposite spectrums neglects the inherent creativity found within the hands of an artisan. Craftsmanship does not merely replicate; it innovates, adapts, and communicates. Each crafted piece, whether a thrown pottery vase or a hand-stitched quilt, carries within it a story, an emotion, a piece of the creator's vision. It is an artistic expression, as worthy of admiration as any painting or sculpture found within the hallowed halls of museums and galleries.

This recognition of the artisan as an artist is not a radical new perspective but a reawakening to the intrinsic value that craftsmanship holds within society. As our culture increasingly values uniqueness and authenticity, the appeal of handcrafted items has surged. This is not merely a trend but a shift towards appreciating the depth of skill, the creativity involved, and the personal touch that comes with artisan-made items.

The marriage of technology and traditional craftsmanship further elevates the artisan's work to new heights. With the advent of digital fabrication technologies, such as 3D printing, and augmented reality, artisans are empowered to explore new forms and express their visions in ways previously unimaginable. The digital realm opens up a world of possibilities, allowing traditional skills to merge with innovative techniques, thus pushing the boundaries of what can be conceived and created.

In educational spheres, there is a growing emphasis on nurturing this blend of artistry and expertise. Schools and institutions that once focused solely on either fine arts or trades are recognising the importance of a curriculum that embodies both. By fostering an environment where students can master the technical skills of a trade while encouraging their creative expression, we are preparing a generation that can truly innovate within their crafts.

The role of social media and online platforms in showcasing craftsmanship cannot be understated. Artisans, with the ease of a few clicks, can now share their creations with a global audience. This visibility not only allows for an appreciation of their work on a broader scale but also positions them as artists in their own right. The stories behind their pieces, the process of their creation, and the personal journey of the artisan become part of the art itself.

As we navigate the challenges and opportunities of the 21st century, the intersection of sustainability and craftsmanship brings another dimension to the artisan's artistry. In a world grappling with environmental crises, the handcrafted stands as a beacon of responsible production. Artisans are at the forefront of the sustainable movement, their practices embodying the principles of eco-friendliness, resourcefulness, and longevity.

This shift towards sustainable and ethically produced items adds layers to the artisan's role as an artist. Their work is not just seen as an object of beauty but as a statement, a choice, an active participation in shaping a better world. It's a form of activism, a canvass for expressing values and inspiring change.

Public perception of craftsmanship and its value in modern society is undergoing a transformation. Events like craft fairs, exhibitions, and festivals play a crucial role in this, offering artisans a platform to showcase their artistry. These gatherings are not just marketplaces but cultural experiences that celebrate the artisan's creativity, skill, and contribution to the arts.

The collaboration between artisans and designers, architects, and other creatives opens up exciting avenues for innovation. Such partnerships highlight the versatility and adaptability of craftspeople, showcasing their ability to engage in dialogues across diverse fields and translate ideas into tangible forms. It's in these collaborations that the full spectrum of the artisan's artistry is displayed, breaking any remaining boundaries between craft and high art.

As the world increasingly values experiences over things, the story of how an object is made becomes almost as important as the object itself. The artisan, as a storyteller, invites consumers into their process, humanising each creation. This connection, fostered through the art of craft, resonates deeply with a society seeking meaning and authenticity.

The recognition of artisans as artists is not merely about semantics but about acknowledging the depth, value, and creativity inherent in craftsmanship. It's about viewing their work through a lens that appreciates the skill, tradition, innovation, and personal expression involved.

Ultimately, the artisan as an artist challenges us to rethink our definitions of art, creativity, and value. In their hands, materials are transformed, traditions are honoured, and innovation is birthed. As we continue to blur the lines between art and craft, we not only elevate the status of the artisan but also enrich our cultural landscape. The future of craftsmanship lies in this recognition, and in embracing the artisan's role as a modern artist, we pave the way for a more thoughtful, sustainable, and beautiful world.

As we move forward, let us celebrate the artisan not just as a skilled craftsperson but as an essential artist of our time. Their work, a blend of tradition and innovation, sustainability, and artistry, demands our respect and admiration. In recognising the value of craftsmanship, we acknowledge its power to inspire, engage, and transform both the individual and society. The artisan as artist is a testament to the enduring human spirit of creation, a bridge between the past and the future, and a vibrant part of our cultural heritage and contemporary life.

Chapter 25:
Preparing for the Future: Skills and Ideas

As we navigate through an era where the fusion of innovation and tradition shapes the landscape of craftsmanship, it's imperative to arm oneself with a mindset geared towards perpetual growth and adaptability. Lifelong learning emerges not just as an option but as a necessity for those keen on carving a niche in this dynamically evolving field. It's about cultivating a voracious curiosity and an unwavering commitment to mastering new techniques, technologies, and methodologies. To stay ahead, one must anticipate market trends, understanding that today's niche interest could burgeon into tomorrow's mainstream demand. This foresight allows craftsmen to align their projects with future needs, ensuring relevance and sustainability in their careers. However, the crux lies not solely in acquiring new skills but in fostering a holistic understanding of how digital advancements can amplify the essence of traditional craftsmanship, creating a new genre of artisans who are as versed in coding or CAD software as they are in the age-old practices of their craft. Thus, preparing for the future demands a balanced amalgamation of skills and ideas, where innovation complements tradition, and adaptability ensures longevity in the ever-evolving narrative of craftsmanship.

Lifelong Learning

In the evolving landscape of craftsmanship, the concept of lifelong learning has taken on a new and exciting dimension. It's no longer about merely honing a singular skill to perfection over decades. Rather, it's about continuously adapting, expanding, and synthesising new skills with traditional expertise to navigate the rapidly changing technological and cultural landscape. This journey of perpetual learning and adaptation is critical for anyone aspiring to excel in modern craftsmanship.

The digital age has ushered in innovations that have significantly altered the way we approach traditional trades. From 3D printing to the integration of augmented reality in design processes, these technological advancements require craftsmen to adopt a mindset of continuous learning. It is essential for craftsmen to not only be skilled with their hands but also be adept with the technology that can enhance the precision, efficiency, and creativity of their work.

Lifelong learning in craftsmanship also involves a deeper understanding of sustainable practices. As the world becomes increasingly aware of environmental issues, the demand for eco-friendly materials and methods in crafting has soared. Artisans today must learn and incorporate these practices into their work to stay relevant and appeal to a market that values sustainability.

Furthermore, the resurgence of the DIY culture and the Maker Movement among Millennials and Generation Z has highlighted the importance of collaborative learning. Participating in online communities, workshops, and maker fairs, provides invaluable opportunities for sharing knowledge and learning new skills from peers. This culture of collaboration encourages artisans to expand their skill-set beyond traditional boundaries.

In addition to technical skills, soft skills play a significant role in the success of modern craftsmen. Skills such as critical thinking, problem-solving, adaptability, and the ability to collaborate effectively are increasingly important. These skills enable craftsmen to navigate the complexities of the modern market, drive innovation, and lead successful projects.

The role of formal education and skill development programmes has also evolved. Trade schools, apprenticeships, and online learning platforms now offer a wide range of courses that combine traditional skills with new technologies. These programmes are designed to equip the new generation of craftsmen with a broad skill-set that responds to the demands of the contemporary market.

Entrepreneurship is another crucial aspect of lifelong learning for craftsmen. The current market landscape offers numerous opportunities for craftsmen to venture into entrepreneurship. Learning about branding, digital marketing, and social media management has become as essential as mastering the craft itself. This knowledge empowers artisans to showcase their work to a global audience and establish successful artisanal businesses.

The value of mental health and well-being in the context of lifelong learning cannot be overstated. Engaging in continuous learning can be a source of joy and fulfilment but can also lead to burnout if not managed properly. It is important for craftsmen to cultivate a balance, ensuring that their passion for learning and creating does not adversely affect their well-being.

Craftsmanship is no longer confined by geographic boundaries. The influence of global cultures on traditional trades provides a rich source of inspiration and learning. By embracing a global perspective, craftsmen can incorporate diverse techniques and styles into their work, further enriching the craft.

The integration of art and craftsmanship presents an additional learning curve. Craftsmen must navigate the nuances of these fields, exploring how they can express their individual identity and narratives through their work. This convergence of art and craft requires not only technical skills but also a deep understanding of the storytelling and emotional connection that art evokes.

To prepare for the future, artisans must stay ahead by anticipating market trends. This involves not only adapting to current trends but also predicting future needs and innovations. Lifelong learning enables craftsmen to remain flexible and visionary, ensuring their relevance in an ever-changing market.

Despite the myriad opportunities that the digital age offers, craftsmen face challenges that require a continuous effort to overcome. Issues such as financial hurdles, the struggle for recognition, and navigating the complex world of intellectual property are all obstacles that craftsmen must learn to navigate effectively.

Lifelong learning also plays a critical role in advocacy and political engagement. Craftsmen must understand the regulatory landscape that impacts their work and engage in advocacy to protect and promote the interests of their trade. This may involve lobbying for changes in policy, engaging with trade organisations, or contributing to discussions on critical issues such as sustainability and inclusion.

The path of lifelong learning is both a personal journey and a communal endeavour. It requires a commitment to personal growth, a willingness to experiment and take risks, and an openness to sharing knowledge and learning from others. This journey enriches the craftsman's practice, ensuring that the tradition of craftsmanship not only survives but thrives in the modern age.

As we move forward, the future of craftsmanship will undoubtedly be shaped by those who embrace lifelong learning. By integrating traditional skills with new technologies, adopting sustainable practices, and fostering a culture of collaboration and innovation, craftsmen can create a future that honours the past while boldly stepping into the new horizons of possibility.

Anticipating Market Trends

In the preceding chapters, we've explored the foundations and evolution of craftsmanship, underlining the critical importance of integrating technology and sustainability. As we navigate further into the future, understanding and anticipating market trends becomes not just beneficial but essential. The ability to foresee shifts within the crafts market can set the groundwork for innovation and longevity in any craftsman's career.

Firstly, it's essential to recognize the role of digital technology in shaping consumer interests and demands. The rise of social media platforms and online marketplaces has not only widened the market for handmade goods but also changed the way consumers discover and interact with artisans. Craftsmen must adapt by not only being present on these platforms but also by engaging with their audience, understanding their preferences, and anticipating their needs.

Environmental concerns are increasingly influencing consumer choices. There's a growing demand for products made from recycled, upcycled, or sustainably sourced materials. This trend is not just a fleeting preference but a shift towards a more eco-conscious society. Craftsmen who incorporate sustainable practices into their work will not only contribute towards a healthier planet but also align with the values of a significant segment of the market.

Another trend shaping the future of craftsmanship is the desire for personalisation and uniqueness. In a world dominated by mass production, the allure of custom-made, bespoke pieces has never been stronger. This longing for individuality presents a golden opportunity for craftsmen to innovate and offer personalised services or products, distinguishing their work in a crowded marketplace.

Moreover, the intersection of art and technology presents new avenues for craftsmen to explore. Techniques such as 3D printing and digital fabrication, when combined with traditional methods, can result in truly innovative creations. Staying ahead of the curve in incorporating these technologies can open up new markets and attract a tech-savvy audience.

Craftsmen must also be aware of the global influences on local markets. With the increasing ease of international trade and online commerce, local markets are being influenced by global trends. Craftsmen should look beyond their localities and consider the global marketplace, adapting their offerings to appeal to a broader audience.

The resurgence of the DIY movement and online tutorials has empowered consumers to take on crafting themselves. This trend can be viewed as a challenge or an opportunity. Craftsmen can capitalise on this movement by offering workshops, DIY kits, or even online classes, thus tapping into the market of aspiring craftsmen and hobbyists.

Understanding demographic shifts is also crucial in anticipating market trends. As Millennials and Generation Z become the predominant consumers, their values and preferences will shape the market. These generations value authenticity, sustainability, and experiences over possessions, indicating a favourable disposition towards handmade and artisan products.

The concept of 'slow living' and mindfulness has been gaining traction, influencing consumer behaviour towards valuing quality over quantity. This trend can be leveraged by craftsmen who focus on creating high-quality, durable items that resonate with the ethos of conscious consumption.

Finally, navigating the economic landscape requires agility and adaptability. Economic fluctuations can impact consumer spending, and being attuned to these changes can help craftsmen make strategic decisions about pricing, marketing, and product development.

To stay ahead, craftsmen must not only keep their fingers on the pulse of current trends but also engage in continuous learning and skill development. This proactive approach ensures resilience and relevance in a rapidly evolving market.

Engaging with other craftsmen and being part of a community can also offer invaluable insights into market trends. Collaborations and exchanges can spark innovation and provide a support network to navigate the complexities of the market together.

The use of data and analytics cannot be overstated in understanding market dynamics. By analysing consumer behaviour, feedback, and trends, craftsmen can make informed decisions to steer their craft in directions that meet the market's demands.

Anticipating market trends is not about predicting the future with certainty but preparing to meet it with flexibility, innovation, and an open mind. The future of craftsmanship lies in the ability to blend tradition with innovation, and embracing market trends is a step towards securing a place in the future landscape of craftsmanship.

In conclusion, anticipating market trends is an integral part of preparing for the future. It involves a constant cycle of learning, adapting, and innovating. Embracing change and being proactive in meeting the market's evolving demands will ensure that the craft and the craftsman remain relevant and cherished for generations to come.

Chapter 26:
A Global Perspective: Craftsmanship Without Borders

In an era of unprecedented global connectivity, the art of craftsmanship finds itself at a unique crossroads, bridging geographical divides and fostering a spirit of collaboration that transcends borders. This chapter explores how craftsmen around the world are breaking away from the siloed practices of the past to engage in vibrant dialogues of exchange and innovation. With the advent of the internet and social media, artisans are no longer confined to local markets; they're connecting, sharing, and learning from each other at a global scale. This intercontinental camaraderie is not just about showcasing techniques and products; it's a potent exchange of philosophies, sustainability practices, and cultural narratives that enrich the spectrum of craftsmanship. By observing the symbiotic relationship between traditional craftsmanship and innovative technologies overseas, one can't help but be inspired by the potential for impactful change. This global perspective underscores a crucial message: the future of craftsmanship hinges not on competing within narrow confines but on embracing a collective ethos that champions knowledge-sharing, mutual respect, and cross-cultural learning. In doing so, it paves the way for a new generation of craftsmen who are as globally minded as they are skilled in their trade, ready to take on the challenges of the future while honouring the legacies of the past.

International Trade and Cooperation

In an era marked by rapid technological advancements and global interconnectedness, the tapestry of traditional craftsmanship weaves an ever-expanding network, transcending geographical boundaries, and cultural divides. The symbiosis of international trade and cooperation offers a fountain of opportunities for craftsmen around the world, enabling an exchange not just of goods, but of ideas, techniques, and inspirations. This chapter endeavours to explore the potent impact of global collaboration on the craftsmanship landscape.

At the heart of this global convergence lies the principle of mutual benefits. Artisans, by engaging in international trade, gain access to broader markets beyond their local confines. This isn't just about selling products but about sharing the soul of their work with an appreciative global audience. It's a dialogue conducted through the universal language of creativity, where the only barriers to comprehension are the limits of imagination.

Digital platforms have emerged as crucial enablers of this phenomenological shift, offering a canvas as vast as the internet itself. Social media, e-commerce sites, and dedicated craft networks have democratised access to global markets, making geographical distance irrelevant. The story of a Peruvian weaver, whose intricate designs find a home in a quaint European boutique, or a Japanese potter whose creations are sought after in North America, is no longer an outlier but a common narrative of success.

However, international trade is not without its challenges. The dichotomy of preserving the authenticity of traditional crafts while meeting the demands of a global market is a tightrope walk for many artisans. Striking a balance between innovation and tradition requires a nuanced understanding of one's craft and the courage to venture into the unknown, adapting and evolving without losing the essence of the artistic heritage.

Moreover, cooperation amongst nations through trade agreements and cultural exchange programmes has significantly bolstered the crafts sector. These initiatives not only facilitate smoother trade relations but also ensure the protection of intellectual property rights, offering a safeguard against exploitation and cultural appropriation.

Sustainability, in the context of international trade, has become a rallying cry for artisans and buyers alike. The global crafts market is increasingly tilting towards eco-friendly and ethically sourced materials, reflecting a collective consciousness towards preserving our planet. This shared mission has paved the way for unprecedented collaborations that underline the role of crafts in leading conversations around sustainability.

Education plays a pivotal role in this global narrative. Skills exchange programmes and workshops conducted across continents serve as a conduit for the transfer of knowledge and techniques, ensuring that traditional crafts not only survive but thrive in the modern era. Young artisans inheriting the mantle from previous generations are thus equipped to navigate the challenges of the contemporary marketplace, armed with an amalgam of traditional wisdom and innovative practices.

The burgeoning trend of collaborative projects across borders has led to a renaissance of sorts in the crafts sector. These partnerships, often between artisans from developing nations and established brands in developed countries, are a testament to the universal appeal and enduring relevance of handmade products. They serve as a powerful reminder that in an age dominated by mass production, the human touch remains irreplaceable.

Success stories of international collaborations abound, inspiring a new wave of craftsmen eager to make their mark on the global stage. These narratives not only highlight the financial gains achievable through such ventures but also the profound sense of fulfilment that comes from witnessing one's cultural legacy resonate with people across the world.

The idea of 'Craftsmanship Without Borders' is not merely a vision for the future but a reality that is taking shape with each passing day. It is a call to action for craftsmen, policymakers, and consumers to foster an environment where creativity knows no bounds, and the art of making connects hearts and minds across continents.

To truly embrace this global perspective, craftsmen must be willing to look beyond the horizons of their local markets and communities. It requires an openness to learn, adapt, and sometimes, unlearn traditional practices that no longer serve their purpose. It's about building bridges, not just between countries but between the past and the future, honouring our heritage while embracing the innovations that define our times.

The role of consumers in this ecosystem cannot be overstated. By choosing to support craftsmen who engage in fair trade practices and uphold the principles of sustainability, consumers can drive significant change. It's a choice that goes beyond the aesthetic value of the products, aligning with deeper values of ethical consumption and global solidarity.

As we forge ahead, the blueprint for a world where craftsmen thrive without borders becomes increasingly clear. It's a world where the spirit of cooperation outweighs competition, where innovation is fuelled by tradition, and where every handcrafted piece tells a story of unity in diversity.

Let us then, embrace this journey with open hearts and minds, ever mindful of the legacy we wish to leave for future generations. The path of international trade and cooperation in craftsmanship is not just about the exchange of goods, but about weaving a tapestry of human connection that spans the globe. It's about crafting a future where art and tradition flourish, hand in hand, in every corner of the world.

Learning from Global Communities

In a world that's more connected than ever before, the opportunity for craftsmen from various backgrounds to exchange ideas, practices, and inspirations is unprecedented. This rich tapestry of global communities offers a boundless well of knowledge for those willing to dip into it. The exchange of craftsmanship techniques, sustainability practices, and aesthetic sensibilities across borders not only enriches the individual artisan but also contributes to a more cohesive global understanding of what it means to create in the 21st century.

Imagine the fusion of Japanese joinery techniques with Scandinavian minimalist design principles. Or think of the vibrant patterns of African textiles being incorporated into the traditional weaving methods of South American artisans. These are not just hypothetical scenarios but real instances of cross-cultural inspiration that have led to innovative products and solutions. This blend of global sensibilities is more than an aesthetic revolution; it's a testament to humanity's shared creativity and ingenuity.

Yet, the learning curve from global communities isn't merely about adopting foreign techniques wholesale. It's about understanding the principles behind them—the philosophy, the approach to materials, and the relationship between the artisan and their craft. It's this deep dive into the essence of different craftsmanship traditions that can yield transformative insights and skills that are both unique and universally resonant.

Consider, for example, the emphasis on sustainability and reverence for nature found in many indigenous crafts traditions around the world. By engaging with these perspectives, modern craftsmen can not only adopt eco-friendly practices but also integrate a philosophical approach to sustainability into their work, viewing their crafts not just as objects but as part of a larger ecological and social fabric.

Moreover, learning from global communities extends beyond traditional craftsmanship. The digital age has brought about a new breed of digital artisans who blend coding, 3D printing, and virtual reality with traditional craftsmanship skills. These tech-savvy creators are forming global networks, sharing knowledge, and collaborating in ways that were unimaginable a few decades ago. They are at the forefront of defining what craftsmanship means in the 21st century.

However, engaging with global communities also presents challenges, such as the risk of cultural appropriation. It's crucial to approach these exchanges with respect, seeking not to exploit but to honour and contribute to the cultures one draws inspiration from. True learning involves a dialogue—a two-way exchange where all parties are acknowledged, respected, and enriched.

Language barriers, too, can pose a challenge, but they're increasingly surmountable thanks to technology. Digital platforms and translation tools have made it easier than ever to connect with and learn from craftspeople around the globe, breaking down walls that once isolated artistic communities.

The benefits of these global exchanges are manifold. They can lead to the revival of dying crafts, giving them a new lease of life by introducing them to new markets and applications. They can help artisans achieve economic sustainability, connecting them with consumers and collaborators worldwide. And importantly, they foster a sense of global community and mutual respect among diverse cultures.

For educators and industry professionals, incorporating this global perspective into learning curriculums and business practices is essential. It's not just about teaching or utilising skills from around the world, but also about nurturing a global mindset—a recognition of the interconnectivity between all craftsmen and the shared challenges and aspirations they face.

The rise of online learning platforms and social media has made accessing knowledge from global communities easier than ever. Whether through formal courses or informal exchanges, these digital spaces are vibrant hubs of learning and inspiration. They demonstrate that, regardless of geographical location, we are all part of a global village of craftsmen.

Young craftsmen, in particular, can benefit tremendously from this open-source approach to learning. By tapping into global communities, they can accelerate their learning process, gain insights into emerging trends and technologies, and connect with mentors and peers from diverse backgrounds. This not only aids their personal and professional development but also prepares them for a career in an increasingly globalised craft market.

The potential for innovation when traditional craftsmanship meets modern technology, informed by a tapestry of global influences, is immense. It's about looking beyond one's immediate environment and embracing a world of possibilities. By doing so, craftsmen can not only push the boundaries of what's possible with their hands and minds but also contribute to a global culture of innovation, sustainability, and shared heritage.

In conclusion, learning from global communities isn't just an option; it's a necessity for any craftsman aspiring to make a mark in the modern world. It's about building bridges—between the old and the new, the local and the global, tradition and innovation. As craftsmen, by embracing this global perspective, we equip ourselves not just with diverse skills and inspirations but with the understanding and empathy necessary to craft a better, more inclusive future for all.

Thus, the journey of learning from global communities is not a solitary one. It's a collective voyage that promises to enrich our crafts, our cultures, and ultimately, our shared humanity. As we look to the future, let's not just carry forward the legacy of our local traditions but also weave into them the threads of global wisdom and creativity, crafting a tapestry that reflects the best of what we, as a global community, can achieve together.

Chapter 27: The Path Forward - Embracing Change and Tradition

As we conclude our exploration into the evolving landscape of craftsmanship, we stand at the cusp of a thrilling age where tradition and innovation intersect, creating unprecedented opportunities for both new and seasoned craftsmen. The journey through the pages of this discourse highlights not only the profound heritage of craftsmanship but also the dynamic shifts propelled by the enthusiasm and ingenuity of millennials and Gen Z. These shifts have set the stage for a future where the artisan's touch merges seamlessly with the digital pulse of technology.

The essence of craftsmanship, with its deep-seated roots in tradition, has demonstrated a remarkable resilience, adapting to the ebbs and flows of societal and technological changes. This resilience reflects a timeless truth: craftsmanship is not about clinging to the past; it's about preserving the core values that define it—quality, sustainability, and a profound connection to the human experience—while embracing the tools and techniques that the future holds.

Indeed, the incorporation of digital technology into the craftsman's toolkit has opened a world of possibilities. From the precision of 3D printing to the immersive experiences offered by virtual and augmented reality, technology has not only expanded the creative horizons but also democratized access to the world of craftsmanship. However, this digital dalliance does not spell the end for traditional methods; rather, it provides a complementary perspective that enhances the inherent value of handmade products.

The challenge, then, for the modern craftsman is not in choosing between tradition and innovation but in finding a harmonious balance that respects the past while boldly stepping into the future. This balance is critical in fostering a sustainable model of craftsmanship that not only preserves our planet through eco-friendly practices but also ensures the longevity and relevance of the trade itself.

Education and skill development represent pivotal elements in this endeavour. Reimagining apprenticeships and trade schools to include modern technology and sustainable practices alongside traditional techniques is imperative. This approach will equip aspiring craftsmen with a diverse skill set that is both broad in its scope and deep in its understanding of the intricacies of the trade.

The evolving job market and the entrepreneurial spirit that characterizes this digital age also offer fertile ground for innovation in the crafts sector. By embracing the opportunities for entrepreneurship and innovation, craftsmen can contribute to a vibrant and dynamic economy that values unique, handmade products and the stories they tell.

This narrative is further enriched by the culture of DIY and the breaking down of stereotypes, particularly concerning the role of women in trades. Online communities and the maker movement have opened doors for collaboration and learning, creating a more inclusive environment that celebrates diversity and fosters creativity.

Social media, too, plays a significant role in modern craftsmanship by providing platforms for branding, marketing, and community-building. Through the power of storytelling, craftsmen can connect with their audience on a personal level, sharing the passion and dedication that goes into every handmade piece.

However, the path forward is not without its challenges. Financial hurdles, the struggle for recognition, navigating bureaucracy, and protecting intellectual property are but a few of the obstacles that young craftsmen may face. Yet, it is through overcoming these challenges that the craftsmanship community will grow stronger, united in their pursuit of excellence and innovation.

Collaborative projects and global collaborations highlight the interconnectedness of the modern world and the potential for craftsmanship to transcend borders. By learning from global communities and embracing a worldwide perspective, craftsmen can draw upon a rich tapestry of techniques, styles, and ideas to enrich their own practices.

Addressing climate change and promoting sustainability are also critical considerations for the future of craftsmanship. By adopting sustainable production methods and contributing to the circular economy, craftsmen can play a pivotal role in shaping a more environmentally conscious society.

Inclusion and accessibility must also be at the forefront of our efforts to evolve the world of crafts. By breaking down barriers and implementing inclusive practices, we ensure that the world of craftsmanship remains open and welcoming to all, regardless of their background or abilities.

The intersection of art and craftsmanship blurs the lines between these two worlds, reinforcing the notion that artisans are indeed artists in their own right. This perspective encourages a greater appreciation for the artistic value of handmade products and the creativity that drives the craft.

Preparing for the future involves a commitment to lifelong learning and an anticipation of market trends. By staying curious, adaptable, and open to new ideas, craftsmen can ensure that their skills remain relevant and their contributions valued in an ever-changing world.

In the end, the path forward for craftsmanship is a tapestry woven from threads of change and tradition, innovation and heritage. By embracing this duality, we pave the way for a future where craftsmanship continues to thrive, enriched by the diversity of its expressions and the depth of its roots. Together, let's step into this future, armed with the tools of our trade and the vision of a world where craftsmanship remains a vibrant and vital part of our cultural and economic landscape.

Appendix A: Resources for Aspiring Craftsmen

In this ever-evolving landscape of craftsmanship, where tradition dances harmoniously with innovation, resources play a pivotal role. The goal is not just to inspire but to equip the next generation of craftsmen with the tools they need to forge their path. What follows is a carefully curated list of resources designed to serve as your compass in the vast sea of craftsmanship possibilities.

Online Learning Platforms

The digital world is rich with platforms offering courses that blend traditional skills with modern technology. Whether you're interested in woodworking, pottery, or digital fabrication, there's something out there for everyone. Websites like Coursera, Udemy, and Skillshare host a plethora of courses that can help elevate your craft to the next level, all from the comfort of your home.

Books and Publications

The Craftsman by Richard Sennett offers a deep dive into the philosophy and ethic behind craftsmanship throughout history and its relevance today.

Handmade: Creative Focus in the Age of Distraction by Gary Rogowski explores the intrinsic value of working with your hands and how it can lead to mastery in any field.

Specialised magazines such as *Make:* and *Fine Woodworking* provide a continuous source of inspiration and practical advice.

Tools & Equipment Suppliers

For craftsmen, the quality of their tools can significantly affect the quality of their work. Research suppliers who not only offer high-quality tools but also sustainable and ethically sourced materials. Local artisan markets and online stores like Etsy can be excellent places to start.

Apprenticeships and Workshops

There is no substitute for hands-on experience. Many master craftsmen offer workshops and apprenticeships, providing invaluable opportunities to learn traditional techniques firsthand. Additionally, institutions and community centres often host classes led by experienced artisans.

Communities and Networks

The journey of craftsmanship can be incredibly rewarding when shared. Joining online forums such as Reddit's r/crafts or specialized Facebook groups can connect you with like-minded individuals. These communities can offer support, answer questions, and provide feedback on your work. Furthermore, local meetups and guilds can present networking opportunities and foster deep connections within the artisan community.

Exhibitions, Fairs, and Festivals

Participating in or attending craft exhibitions, fairs, and festivals can offer a broad perspective on the current trends, techniques, and materials in various trades. Such events can also be a fantastic way to showcase your work and connect with potential mentors, collaborators, or customers.

Innovation and Technology Resources

For those aspiring to blend craftsmanship with technology, exploring resources related to 3D printing, laser cutting, and CNC machining can be incredibly beneficial. Websites such as Instructables or platforms like Maker's Muse on YouTube offer tutorials and project ideas that can spark your creativity and technical skills.

Final Thoughts

Embarking on the path of craftsmanship is a journey of continuous learning and discovery. The resources listed here are just the beginning. As you delve deeper, you'll find your unique blend of tradition and innovation, channeling your passion into creations that resonate with the world. Remember, every master was once a beginner, and with perseverance, creativity, and the right resources, you too can leave a mark on the age-old tapestry of craftsmanship.

Chapter 28: Acknowledgements

As we bring the journey of this book to a close, it's imperative to pause and reflect on the multitude of individuals and experiences that have made this endeavour not just possible, but profoundly enriching. Crafting this narrative has been a voyage of discovery, inspiration, and unending learning, attributable in no small part to a wide array of contributors whose insights, stories, and passion have infused these pages.

First and foremost, I'd like to express my heartfelt gratitude to the artisans and craftsmen who have generously shared their stories, their struggles, and their successes with me. It's their dedication to their crafts that stands at the very heart of this book. Their willingness to open up their workshops, share their processes, and divulge the intricacies of their trades has been nothing short of inspiring.

To the educators and mentors across the globe, who tirelessly work to preserve the essence of craftsmanship and pass it on to the next generation, your role cannot be overstated. Your insights into the integration of technology with traditional skills have been invaluable, not just for the creation of this book, but more importantly, for the countless students whom you inspire daily.

A special note of thanks must go to the team behind the scenes. To the editors, who meticulously combed through drafts, ensuring clarity and coherence, your patience and expertise have been pivotal. The designers and illustrators deserve immense credit too, for bringing the essence of craftsmanship to life with their visuals, making this not just a book, but a piece of art in itself.

To the researchers and academics who have dedicated their careers to studying the evolution of trades and technology, thank you for your foundational work, which has been a cornerstone of this book. Your papers, articles, and discussions have not only informed its content but have also stirred debates and thoughts that enriched it manifold.

I extend my gratitude to the organisations and community groups that support craftsmen across different stages of their careers. Your initiatives and programmes provide the ecosystem in which traditional trades can flourish alongside modern innovation. The work you do in fostering networks, championing sustainability, and promoting inclusivity has been both motivational and instructional.

Gratitude is also due to my peers and colleagues, whose critiques and suggestions have been invaluable. Your perspective has often challenged me to think differently and dig deeper, making this work richer and more nuanced. The conversations, debates, and sometimes heated discussions we've had have all contributed to the depth of this book.

To those in the tech industry, who are innovating at the crossroads of tradition and technology, thank you for demonstrating what the future of craftsmanship can look like. Your work in 3D printing, virtual reality, and the Internet of Things has not only been featured in these pages but has also sparked the imagination of what's possible.

I am immensely thankful for the online communities, forums, and social media platforms that have allowed me to engage with craftsmen, enthusiasts, and readers from all over the world. Your feedback, questions, and encouragement have been instrumental in shaping this book. The sense of community and shared passion has been a constant source of inspiration.

To the young individuals who stand at the threshold of their careers, pondering a path in craftsmanship, your curiosity, enthusiasm, and willingness to embrace both tradition and innovation have been the driving force behind this book. It's my sincere hope that these pages encourage you to pursue your passion and make your mark in the evolving landscape of craftsmanship.

My family deserves a special mention for their unwavering support and patience throughout the process of writing this book. Your love, encouragement, and sometimes, sacrifice, have not gone unnoticed. It's your belief in the importance of this work that has sustained me during challenging times.

Finally, to the readers who have embarked on this journey with me, your engagement and interest are what ultimately give this book its purpose. It's been written with the aim of not just informing but also inspiring and provoking thought. Your willingness to explore the fusion of tradition and technology in craftsmanship is what will carry this conversation into the future.

In conclusion, crafting a book is much like crafting any piece of art; it's a process that demands patience, persistence, and passion. The many individuals who have contributed their expertise, stories, and support have been indispensable to this process. As we look towards the future of craftsmanship, let's carry forward the spirit of innovation, respect for tradition, and the dedication to mastering one's craft that has been the hallmark of the artisans featured in these pages. Here's to embracing change, cherishing tradition, and crafting a future where both can thrive together.

Thank you, one and all, for being part of this journey.

www.ingramcontent.com/pod-product-compliance
Lightning Source LLC
Chambersburg PA
CBHW030005190526
45157CB00014B/430

* 9 7 8 1 4 5 6 6 4 8 7 5 6 *